LANGHANS, ROBERT W
A GROWTH CHAMBER MANUAL: ENVIR
000315949

QK731.L27

KU-405-525

WITHDRAWN FROM STOCK
The University of Liverpool

A Growth Chamber Manual

Environmental Control for Plants

A GROWTH CHAMBER MANUAL

Environmental
Control
for Plants

Edited by

Robert W. Langhans

UNIVERSITY LIBRARY LIVERPOOL

COMSTOCK PUBLISHING ASSOCIATES

A division of Cornell University Press
Ithaca and London

Copyright © 1978 by Cornell University

All rights reserved. Except for brief quotations in a review, this book, or parts thereof, must not be reproduced in any form without permission in writing from the publisher. For information address Cornell University Press, 124 Roberts Place, Ithaca, New York, 14850.

First published 1978 by Cornell University Press.

Published in the United Kingdom by Cornell University Press Ltd., 2-4 Brook Street, London W1Y 1AA.

International Standard Book Number 0-8014-1169-6

Library of Congress Catalog Card Number 77-90906

Printed in the United States of America

Librarians: Library of Congress cataloging information appears on the last page of the book.

Preface

This book introduces growth chambers to the new user and serves as a reference for all users—technicians, researchers, and students. The Committee on Growth Chamber Environments of the American Society for Horticultural Science has pointed out the need for a book that explains how a growth chamber works, how to grow plants in a chamber, and how to avoid many of the pitfalls and problems associated with growth chamber operation. This book is intended to fill that need.

Growth chambers have proliferated like well-grown callus tissue cultures, and almost all plant science departments from those in secondary schools to those in pure research institutions have at least one growth chamber. There are thousands of chambers in the United States and many more all over the world, ranging from the very simple to the sophisticated.

While manufacturers of growth chambers always give installation instructions and engineering specifications, their brochures unfortunately supply limited information on proper management and operation. Typically, technical information on such topics as temperature variability, light intensity, and relative humidity range is evaluated under ideal conditions in the manufacturer's laboratory, usually without plant material in the chamber. Information on preparing specifications for growth chambers, measuring and controlling the environmental conditions, growing the plants, controlling diseases and insects, maintaining the chamber, and determining the experimental designs is difficult or impossible to find. This handbook brings together in one volume conversion tables for units of measurement and specification data for chambers and for the equipment needed to regulate environmental conditions within them.

The lack of information on management and operation has led to uncertain and greatly variable experimental results. Editors

5

552830

of most scientific journals now require an author to report the specific conditions found in the chamber; no longer is it possible to say, "The plants were grown in a growth chamber." For this reason the revised guidelines prepared by the Committee on Growth Chamber Environments for reporting studies in controlled environment chambers are included as Chapter 14. The guidelines recommend that reports of studies conducted in growth chambers identify by trade name the measuring and recording devices used; nevertheless, mention of a trade name in this book does not constitute an endorsement, guarantee, or warranty of the product by the authors and does not imply its approval to the exclusion of other products that may be suitable.

The contributors have had a great deal of experience with growth chambers, and in most cases the particular subject about which they write has been the focus of their research. They have provided references at the end of each chapter to help the reader investigate the subject more thoroughly. We hope that this handbook will give researchers the information necessary to make their growth chambers accurate and useful research tools.

The cooperation of the authors is gratefully acknowledged. Special appreciation is extended to Marian Rollins for typing the manuscript and to my wife, Virginia Langhans, for preparing the illustrations.

R. W. LANGHANS

Ithaca, New York

Contents

Figures

Tables

Contributors

William A. Bailey, Agricultural Engineer, U.S. Department of Agriculture, Agricultural Research Service, Beltsville, Maryland.

Wade L. Berry, Specialist, Laboratory of Nuclear Medicine and Radiation Biology, University of California, Los Angeles.

Geoffrey Burdge, Electrical Engineer, formerly with U.S. Department of Agriculture, Agricultural Research Service, Beltsville, Maryland.

William A. Dungey, Engineering Technician, U.S. Department of Agriculture, Agricultural Research Service, Beltsville, Maryland.

P. Allen Hammer, Professor, Department of Horticulture, Purdue University.

Richard H. Hodgson, Plant Physiologist, Metabolism and Radiation Research Laboratory, North Central Region, U.S. Department of Agriculture, Agricultural Research Service, Fargo, North Dakota.

R. Kenneth Horst, Professor, Department of Plant Pathology, Cornell University.

Herschel H. Klueter, Agricultural Engineer, U.S. Department of Agriculture, Agricultural Research Service, Beltsville, Maryland.

Donald T. Krizek, Plant Physiologist, Plant Stress Laboratory, U.S. Department of Agriculture, Agricultural Research Service, Beltsville, Maryland.

Robert W. Langhans, Professor, Department of Floriculture, Cornell University.

J. Craig McFarlane, Plant Physiologist, Environmental Protection Agency, Environmental Monitoring and Support Laboratory, Las Vegas, Nevada.

Douglas P. Ormrod, Professor, Department of Horticultural Science, University of Guelph.

Theodore W. Tibbitts, Professor, Department of Horticulture, University of Wisconsin, Madison.

A Growth Chamber Manual

Environmental Control for Plants

Chapter **1** J. CRAIG MCFARLANE

Light

Light, in growth chambers as in the field, is the sole source of energy for plant growth and development. The effect of light on plants has been the subject of countless studies in photosynthesis, photomorphogenesis, and bioenergetics. Our intent is not to provide extensive instruction on these topics, but to discuss light production and measurement in sufficient detail to allow a fuller utilization and understanding of plant growth chambers. Only a brief overview of the basic concepts of light reactions on plants will be included.

Characteristics of Light

Light is defined as transverse electromagnetic radiation with appropriate wavelength to stimulate the sensation of vision in the normal human eye. Thus, the word light is strictly defined only in terms of a human physiological response and it is conditioned by terms such as normal or standard human vision. Most measuring devices are influenced by radiation beyond the limits of human perception and light sources typically produce radiation outside the spectral region called light. Also some plant responses are mediated by nonvisible radiation. Thus in this text as in the common vernacular a modified definition will be adopted. We will use the word "light" to mean all visible radiation plus the regions of infrared and ultraviolet. Figure 1.1 shows the spectrum of radiant energy, which includes light as only a small segment.

In 1901, Max Planck reluctantly published one of the most important of scientific concepts. He hypothesized that radiation was discontinuous and could only be produced with discrete en-

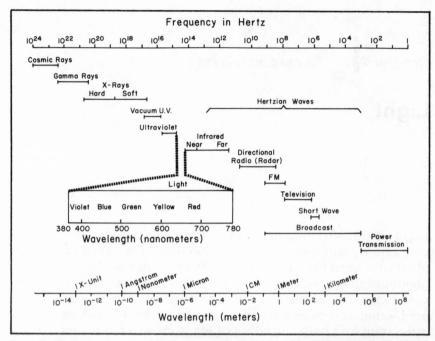

Fig. 1.1. The radiant energy (electromagnetic) spectrum. (*IES Lighting Handbook* 1972.)

ergy values ($\lambda = hv$), where h was constant, v the frequency of light, and λ the photon energy in terms of ergs per second. Four years later, Albert Einstein showed that energy was only absorbed in discrete steps and confirmed that radiation was composed of energy packets, called quanta or photons. In describing phenomena, such as photosynthesis, that depend on the absorption of light, quanta are used because they define the energy absorbed. Since h is a constant, the energy (λ) is directly proportional to the frequency (v). Frequency is reciprocally proportional to wavelength and the energy of light increases with decreasing wavelength. Thus, a quantum of blue light has more energy than one of red light. Figure 1.2 shows quantum energy as a function of wavelength from 300 to 800 nm. In order to cause a photochemical reaction, a photon must be absorbed in such a way that its energy is transferred to an electron of a molecule, atom, or ion. This electron with the additional energy changes orbit, becoming excited and capable of reacting in various ways. This phenomenon is the origin of reactions such as photosynthesis, flower initiation, and photonastic movements. It also accounts for reactions that allow light measurement.

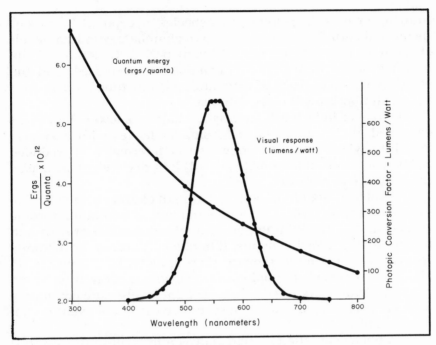

Fig. 1.2. Quantum energy of photons and visual response

Light Reactions in Plants

Primarily, light is the energy source for the photosynthetic process, which determines the carbon fixation rate and thus controls the rate of plant growth. Light also provides a stimulus that initiates a number of other reactions vital to plant development.

Plant growth is a complicated expression of many interacting factors, including light intensity, quality, and duration. Various morphological, physiological, and biochemical characteristics peculiar to different species determine their diverse response to different light conditions. For instance, some slow-growing foliage plants will succeed with as little as 1 to 5 nano-Einsteins* per square centimeter per second (nE cm^{-2} s^{-1}) for an 8-hour period. The maximum level for plants with C$_3$ photosynthetic systems is commonly about 32 nE cm^{-2} s^{-1} for a 16-hour period (Tibbitts 1977). For high light plants, particularly those with C$_4$ photosynthetic systems, a light level in excess of 50 nE cm^{-2} s^{-1} for a 16-hour period is required to maximize growth. Vigorous fieldlike

*One nano-Einstein = 6.02 $\times$ 10^4 photons.

growth, for most high light plant species, exceptions being soybeans and corn, is obtained in growth chambers with light levels of 25 to 35 nE cm^{-2} s^{-1} for a 16-hour period. Generally, if the duration of the light period is extended or decreased, and the intensity varied so that intensity multiplied by time is a constant value, similar growth rates will result.

The use of light levels in growth chambers lower than those described above may lead to (a) smaller leaves with a greater length to width ratio, (b) elongation of internodes, (c) reduced concentration of chlorophyll, and (d) less dry weight at a particular stage of maturity.

In contrast, high light levels in growth chambers may lead to increased anthocyanin production and purple coloration in some species and to the stimulation of auxillary branch growth and proliferation of growing points. If light levels exceed the amounts required for maximum growth, there may be bleaching due to photodestruction of chlorophyll. This bleaching is rarely observed in growth chambers since excessive light levels more commonly heat the leaves so that water loss, desiccation, and eventual necrosis result.

The photosynthetic reaction is driven by light in the spectral region between 400 and 700 nm, although photons of radiation at all wavelengths within this range are not equally effective in producing photosynthesis in intact plants. For wavelengths between 500 and 600 nm efficiency decreases slightly, between 10 and 25 percent for different plant species. Light in this region is not absorbed well by chlorophyll but instead is reflected, giving plants the characteristic green appearance. Studies of light reactions in extracted chlorophyll have led some to the false conclusion that photosynthetic efficiency is proportional to the chlorophyll absorbance spectrum. Light absorbed by other pigments, primarily the carotenoids and the phycoblins, also results in photosynthesis when the absorbed energy is transferred to chlorophyll. Photosynthetic or plant action spectra are complex summations of the responses of many pigments and should not be confused with the chlorophyll absorption spectra.

Several responses of plants, such as germination, flowering, and phototaxic movements, result from the mere presence of light and are not influenced greatly by its intensity, provided that certain minimum levels are exceeded. The intensity necessary for some of these responses is the level of moonlight (0.01 nE cm^{-2}

s^{-1}) but they are more commonly controlled by levels that exceed 1.0 to 1.5 nE cm^{-2} s^{-1}. In growth chambers, white light from fluorescent lamps or high intensity discharge lamps provides the control for most of these reactions. Monochromatic light studies have demonstrated that most photostimulus reactions are controlled by wavelengths either in the blue region (400 to 460 nm) or in the red and far-red regions between 650 and 800 nm. The latter reactions between 650 and 800 nm have been found to be controlled or partially controlled by the phytochrome hormones. Narrow band radiation in the red at 660 nm and in the far-red with a peak at 740 nm can be demonstrated to be the controlling wavelengths. Although it would appear that investigators should be very concerned with the balance between red and far-red radiation there is very little data to demonstrate either a requirement for a balance or the necessity of a period of only red or far-red radiation at the initiation or conclusion of the light period. The only presently known examples are for certain photonastic movements. Five photostimulus reactions are important in growth chambers studies.

Germination. A few seeds require light for germination, particularly native plant species that have not been selected for general crop use. This requirement can be met by planting within the top 5 mm of the soil surface and providing light during germination.

Hypocotyl. The hypocotyl of some species straightens as the emerging seedling contacts light.

Flowering. The initiation of flowering in most plants is independent of day length, although in some species critical periods of light and dark are required. The photoperiodism of some species divides them into two groups, short and long day plants. Short day plants generally show a response (flowering) when the daylength is less than the critical, whereas the long day plants show a response (flowering) when the daylength is longer than the critical. The critical daylength is that photoperiod above or below which the reaction occurs. The distinctions are imprecise since most plants change their response to daylength with age and environmental factors such as temperature or light intensity. Some species require a period of long days followed by short days, and some the reverse, to produce the maximum response.

Photonastic movements. Certain movements of plant tissue are initiated by light, including opening and closing of flowers, up and down movement of leaves, and twining of tendrils and stems. Research indicates that short-period movements of bean leaves (30 minutes to 2 hours) and twining of dodder stems require far-red radiation. In growth chambers this requirement can be met with 20 percent of the input wattage provided by incandescent lamps and 80 percent by cool-white fluorescent lamps.

Stem elongation. The internodes of plants will be reduced in length if there is an excess of radiation in the blue wavelengths. Thus, a balance between blue and red light is required.

Measurement of Light

Two systems of light measurement have been developed—photometric and radiometric. The values of these two measurement systems can only be equated by careful consideration of complex interrelationships that depend on the sensitivity of various light receptors.

Photometry deals specifically with the measurement of light seen by the "standard human eye." Like most other light receptors, the eye does not respond to all wavelengths of light with equal efficiency. Efficiency peaks at approximately 550 nm (yellow) and falls off in the blue at 400 nm and in the red at 700 nm (Fig. 1.2). Thus, it takes more light energy in the blue region than in the yellow region to cause the same quantitative response. The efficiencies of zero for ultraviolet and infrared radiation indicate that light in these regions is not visible to the normal human eye.

Radiometric measurements describe incident radiation in absolute energy terms, which are not associated with the efficiency of a particular receptor. The important difference between the two measurement systems is that in the photometric terminology all radiated energy is not regarded equally. Photometric techniques measure the visual sensation caused by light rather than the radiant energy.

Flux densities (power per unit area) are the most useful and the most commonly used indicators of light in growth chambers. Flux describes the total power emitted from a source, or incident on a surface. Illuminance and irradiance are photometric and radiometric terms that define the flux density. In photometric terms, the units are Lux (Lumens $\cdot$ m^{-2}) or foot candles (Lumens $\cdot$ ft^{-2}). In radiometry the most valuable units for growth chamber studies are milliwatts per square centimeter (mW $\cdot$ cm^{-2}).

It is obvious that different light absorbing pigments in plants have different absorption and action spectra. Therefore, various plant functions respond to different spectral bands of light with different efficiencies. It is also obvious that plant pigments are not similar to the pigments responsible for human vision. Therefore, measurement of light for plant studies is a problem not solved by the concept of photometry. Historically plant researchers have had at their disposal only photometric instruments and have used them as an index of light available to plants. Photometric values are weighted as shown in Figure 1.2, and although the weighting is based on a receptor completely foreign to plants, the rationale for using this system has been that photometric values were closer to plant response values than the total energy values of radiometric determinations. Thus, even if unsatisfactory, photometric values were assumed to be better than none. This assumption has unfortunately established foot candles and lux measurements as acceptable values in plant studies.

The deficiencies of existing light measurement techniques have drawn the attention of many critics (Norris 1968, Tyler 1973, Shibles 1976), and efforts to devise a better system have been undertaken by several authors. The concept of physiologically available irradiation (PAI) was suggested in 1953 by the Committee on Plant Irradiation of the Nederlandse Stichting voor Verlichtingskunde. Under this concept, energy (mW · cm^{-2}) is measured in five spectral bands that correspond to major plant response spectra. Confusion as to how this information would be used or interpreted has caused this concept to be poorly accepted. A modification of the PAI system, photosynthetically active radiation (PAR), was proposed by Federer and Tanner (1966). It is a measurement of the incident quanta between 400 and 700 nm and is typically expressed in terms of nano-Einstein per square centimeter per second (nE cm^{-2} s^{-1}). The Crop Science Society of America has recently recommended (Shibles 1976) that this unit be officially designated as photosynthetic photon flux density (PPFD). They also recommended that the irradiant flux density between 400 and 700 nm be termed photosynthetic irradiance (PI). Also, McCree (1966) suggested the use of a plant watt, which he defined as 107 ergs per second in the region between 400 and 700 nm.

There is apparent agreement that the spectral region between 400 and 700 nm is the most important. However, as evidenced from the many different units and measurement criteria suggested, confusion still pervades the literature. Since photosynthesis is the primary plant response exhibiting control over the growth rate and general development of plants and since photosynthesis responds somewhat proportionately to PAR, measurement of quantum flux density has been recently emphasized. Nevertheless, reliance on PAR measurements alone ignores the fact that each plant response is wavelength dependent. Therefore, the main disadvantage of this measurement method, as of many others, is that it does not describe incident light sufficiently. Only one measurement is made (intensity or flux density) and spectral measurements are generally ignored.

Measurement and Reporting

How can radiant energy be measured and results expressed in a meaningful way? The most appropriate method would be one that measures and reports both the spectrum and the radiant flux. This method is easier to describe than to perform. Norris (1968) reviewed measurement techniques and concluded that "spectroradiometric analysis appears to offer the only real solution to the problem of evaluating radiation sources." His review discussed the advantages and disadvantages of many available spectroradiometers and is a good source of information about them. Unfortunately, spectroradiometers are expensive and, therefore, not readily available. Also there are no simple correlations between the spectral patterns of incident light and growth or plant development. Most relationships are complex, a condition created by the peculiar nature of light and the interrelationships among plant functions. Thus, the investigator who is searching for an understanding of light-dependent reactions should endeavor to obtain measurements of both flux and spectra.

Quantum measurements correlate fairly well with photosynthetic rates and with total biomass accumulation. For this reason PAR measurements are becoming the most accepted measurement units. Instruments are available that are filtered to approximate the true quantum flux and that read out in PPFD units (nE cm^{-2} s^{-1}) between 400 and 700 nm. These instruments should be used whenever possible.

Less critical studies may be accomplished with less critical measurement. But since all science is based on the concept of repeatability, experiments conducted in growth chambers should include sufficient data to allow the recreation of experimental conditions. Certain guidelines may be applied to measurement techniques which will allow this accuracy.

A description of the lamps used is of utmost value. Manufacturers publish lamp spectra under known conditions, but as the lamps age or the temperature changes there are some spectral shifts. Thus, the following information should be provided when possible. Lamp description should include manufacturer, type, input wattage, temperature of lamp bank, number and average age of bulbs, rated voltage, and line voltage. Reported values should be accompanied by sensor type, model, and manufacturer; calibration procedure if applicable; and sensitive wave band.

When a combination of lamps is used, independent measurements of each type should be made. A valuable check on the linearity of a sensor is to mathematically sum the independently measured values for each lamp type and compare the total with the value obtained by the sensor when all lamps are illuminated. Remember that fluorescent lamps require about 2 hours to stabilize. If a chamber has a combination of fluorescent and incandescent lamps, the best procedure for light measurement would be as follows: at least two hours after lamps have been on (1) measure light intensity of the combination; (2) turn out incandescent lamps and measure fluorescent lamps alone; (3) turn on incandescent lamps alone, wait about 5 minutes, and make final measurement. The sensor should not be moved between these readings. It is desirable to have more than one sensor response to evaluate chamber lighting. Especially valuable is a measurement of infrared radiation, which, with the data from another sensor, allows an evaluation of spectral shifts in incandescent lamps caused by voltage changes and aging of filaments. Preferred measurements are those of quanta or radiant flux between 400 and 700 nm in combination with a narrow infrared band somewhere between 700 and 800 nm. Despite the deficiencies of photometers in plant growth chamber research they have some legitimate uses, such as the determination of light variation in a chamber. Photometers can also be used to determine absolute incident energy if they are calibrated against a radiometer for a particular light source. When this technique is used, the investigator must be cautioned against thinking that calibration values are applicable to any other light source. If the relative spectrum of a new source is different or the spectrum varies from one point to another in the chamber, obviously the total energy is different. When light sources are known, conversion factors can be determined to render absolute energy data from illumination measurements. The conversion factors calculated by Gaastra (1959) and Bernier (1962) as compiled by Bickford and Dunn (1972) and those reported by Klein (1973) and Tibbitts et al. (1976) are useful in making these estimates.

Determining where and how to measure radiation in plant growth chambers is not always easy. Several concepts are important in making these decisions. The angle of incident radiation to the detector can affect measured flux at the surface. If a detector is covered with a protective glass shield, part of the incident light will be reflected. If reflected, it will not be measured and the flux density will be underestimated. Reflection may account for errors as high as 25 percent depending on the amount of light impinging at steep angles. Correction for reflectance can be accomplished by covering the detector with a white plastic diffusing cover or opal glass. The Lambert cosine law describes the relationship between flux and incident angle: when light strikes a detector at an angle, the value of irradiant flux should equal the value that would occur if the detector were perpendicular to the rays, multiplied by the cosine of the angle. This law is called a cosine correction and, although misnamed, the diffusing cover is called a cosine corrector.

Radiant flux decreases with increasing distance from the source. A decision about the position of light measurement is, therefore, important. Measurement taken at the top of a plant canopy can cause confusion, because plants grow closer to the light source and are eventually in a higher flux field. Some chambers and growth rooms are constructed so that either plant level or light height can be varied. This flexibility is advantageous in tests where radiant flux must be critically maintained at the top of the plant canopy over a period of growth. This technique can also be used to compensate for changes in lamp intensity, which decreases with lamp age. Unless stringent requirements are demanded by the experimental design, radiant flux should be measured either at the top of the plant canopy or at a constant point throughout a study. Regardless of which system is used, such details must be provided when reporting the study.

Radiant flux within most chambers is not uniform. Hammer and Langhans (1972) considered light and other variables in chambers and suggested the use of a complete block randomized statistical design to minimize confusion of results. They observed decreased light intensity in the corners and edges of their chamber and suggested the use of guard plants to reduce the edge effect. When reporting light measurements certain information regarding variability should be included: height of measurement in relation to height of plant canopy, variation within test area, variation with time (see discussion of fluorescent lamps below).

An interesting and valuable exercise is to determine light distribution patterns in your growth chamber. Remember that light intensity varies with the lamp temperature. Consider the cooling system cycles, especially if the light cap has a refrigerated cooling system. On a strip chart, record the light intensity in the center of the chamber from the start of a light period of at least two hours. This plot will reveal the changes that occur during warm up and will allow evaluation of cycling coincident with temperature fluctuation in the light bank. (Figure 1.3 shows the patterns of light intensity in two of the author's chambers measured in the above manner.) Next position empty pots or some other marker on

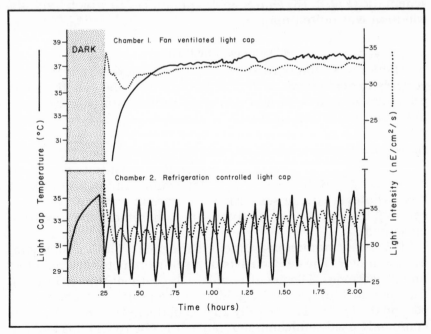

Fig. 1.3. The effect of light cap temperature on light intensity in two growth chambers

the rack or bench in a regular grid. The number depends on your experimental design and the chamber size. Place your sensor on each pot and record the light value. If intense cycling occurs it may be necessary to allow the cooling system to make one complete cycle at each measuring station. Then select a common point—top, bottom, or median of the cycle—as your intensity value. Plots of this data will be useful for the design and evaluation of experiments. Figures 1.4 A and B show the variability in two of the author's reach-in growth chambers.

The data revealed a gradient of light intensity of about 10 percent from the left to the right side of one chamber. Without this evaluation, problems such as the one revealed in Figure 1.4A might have been overlooked. After the temperature gradients in the light cap were evaluated and light measurements were made with and without plants in the chamber, I discovered that there were differences in the reflectance of the chamber walls. In this particular chamber the air moves horizontally through the side walls, and differences in the hole sizes caused this pattern. Simple modifications have diminished this problem. The second schematic (Fig. 1.4B) shows another chamber in which light distribution was quite uniform.

Measurement Instruments

There are so many instruments on the market that it is impossible to evaluate each device. We will describe only the basic principles of detection and point out some features considered important regarding each type.

Photoelectric Detectors

All photoelectric detectors have limited sensitivity and must be corrected with filters if used to determine luminous or radiant flux. Whatever type of sensor is used, care must be taken in discerning the spectral sensitivity and, therefore, the response measured.

The most common instrument is the photovoltaic photometer or light meter. The cell consists of a semiconductor between two metal electrodes. The negative electrode (cathode) on top is a thin transparent film, generally gold, and the positive electrode (anode) on the bottom is generally an iron plate. Selenium is the most commonly used semiconductor since it is cheap and has a sensitivity similar to that of the human eye. Silicon is better in that it does not fatigue as readily, has a higher current, and has less internal resistance than selenium. When exposed to light, an electromotive force (emf) is generated between the cathode and the semiconductor that is proportional to the light intensity. Although the response of photovoltaic cells is maximum for yellow-green light, before being marketed as foot candle meters they must be fitted with special filters to adjust their sensitivity to match the human eye.

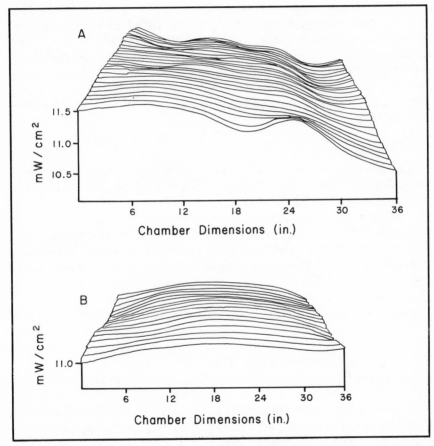

Fig. 1.4. Perspective of light intensity (mW • cm^{-2} from 400 to 700 nm) viewed from the door side of two growth chambers. These presentations were computer drawn from data points taken on a regular grid with 6-inch centers. (A) Differences in reflectance from side walls in this 16 ft^2 chamber created the gradient from left to right. (B) More uniform light distribution occurred in another commercially available 12 ft^2 chamber.

Photoconductive detectors measure the change of resistance caused by light on certain semiconductor materials. They require a stable current source and generally utilize lead sulfide, lead selenide, lead telluride, or cadmium sulfide as the sensitive material. Photoconductive cells with a combination of filters can measure quantum energy. When the filters exclude light outside of the 400–700 nm range, photoconductive cells indicate photosynthetic photon flux density (PPFD).

Photoemissive detectors are gas filled or evacuated tubes containing photosensitive cathodes. When light is absorbed by the cathode, electrons are emitted. An electrical potential causes these electrons to migrate to the anode and produce a current which is proportional to light intensity. Photoemissive detectors are typically used in spectrophotometers and a complicated modification of this device is called a photomultiplier. By using selected filters, they can be used to measure either luminous or radiant flux.

Thermodetectors

Incident radiation absorbed by a black body is reduced to heat and the heat produced is proportional to the energy absorbed. A measurement of the heat provides an unbiased determination of total energy. Thermoelectric detectors typically are thermopiles or bolometers. Thermopiles are made of several thermocouples mounted in series arranged so that the hot junction is in contact with a black absorbing material and thus is heated when radiation is absorbed. Thermopiles are an excellent primary detecting device and are generally used as standardization equipment in calibrating other detectors. They are generally costly, subject to loss of calibration if not handled with care, and often difficult to employ correctly. For instance, some thermopiles have a very narrow angle for interception of incident radiation. In a growth chamber such a device pointed at the light cap may view only a few square centimeters at each setting. Other thermopiles are very sensitive to being moved. One thermopile that has become a standard in meteorological measurements is termed a pyroheliometer. This instrument has a flat surface and is suited to growth chamber application. The sensitivity of any sensor is limited by covering materials such as plastic or glass, which are not totally transparent to all wavelengths of radiation.

Bolometers are resistance thermometers made of black absorbing elements that change temperature with changing incident radiation. The resistance forms one leg of a wheatstone bridge. Their response is linear over a usable range and they are highly accurate. They are not generally used in growth chambers because they are not as portable as thermopiles and they are expensive.

Integrating Instruments

In some situations the investigator may desire to integrate radiant energy over a period of time. Several electronic integrating instruments are available with appropriate filters and scaling to obtain integrated measurements of radiation as light, energy, or quanta.

Chamber Design

Several design features are important in maintaining light intensity and uniformity in growth chambers. The effect of light cap barriers for temperature control of lamps and the type of reflective surface of the chamber walls are salient considerations.

Light Cap Barriers

The merits of a separate temperature control for the light cap and of a transparent barrier between the lamps and the plant area are threefold: (1) they prevent excessive heat in the plant compartment, (2) they maintain uniform light output, and (3) they filter out ultraviolet radiation. The heat produced by the lamps must be removed and some means is required to regulate lamp temperature. Where the lamps are separated from the plant area, it is important that proper temperature control be provided. Changes in temperature cause light intensity to increase or decrease about 1 to 3 percent per degree centigrade (Fig. 1.3). Chambers without barriers may have significant variations in lamp output when experiments are run at widely different temperatures. For instance, the output of fluorescent lamps in an experiment at 15°C may be only 70 percent of the lamp output for a similar experiment conducted at 25°C. Similarly variations can occur when a barrier is used if the temperature of the room air utilized to cool the lamps varies greatly at different times of the day or at different seasons. Chambers that have separate temperature controlled lamp compartments may also suffer from light irregularities. Since the heat generated is large and the air volume is generally small, temperature cycling in the lamp cap is often sufficient to produce significant irradiance cycling. Therefore, each user must evaluate his own chamber to know what conditions exist.

Reflective Surfaces

Chamber manufacturers generally offer a variety of different materials for walls and ceilings. Specular materials, either aluminum coated with mylar or stainless steel, have been extensively used. Some chambers are constructed with gloss or flat white sides and ceilings and some even have black sides to eliminate reflected light from the lower parts of the plants. A collection of reflectance curves was presented by Hollaender (1956) and represented by Bickford and Dunn (1972). This information is valuable for choosing wall and ceiling materials. Tests show that specular aluminum yielded approximately 85 percent reflectance over a broad spectrum including the important wavelengths between 300 and 800 nm. Magnesium oxide paint had the highest reflectance with the least dependence on spectrum but it is not routinely used because it requires several coats to cover and is difficult to maintain. Tests of various materials in an actual plant growth chamber were conducted by the Bioclimatic Laboratory Development Project at Cornell University in 1961 (Davis and Dimock). Their data showed that specular ceilings and walls gave the highest light values in the chambers. The difference between specular and diffuse ceilings was found to decrease as the lamps were moved closer together. White diffuse walls gave more uniform light distribution in the chamber, but lower intensity. One must therefore choose between maximum uniformity and maximum intensity. Most users sacrifice some uniformity for greater intensity. Uniformity can be partially manipulated by placing the newest bulbs near the chamber sides and moving them toward the center when replacing old ones. Shading can also be used within the chamber to increase uniformity but this requires great patience and considerable skill. In order to maintain maximum light reflectance it is important to keep chamber walls free of algae growth, water spots, and other residues.

Light Sources

Light intensity as well as light quality can be controlled and altered in plant growth chambers to satisfy the needs of most researchers. In most instances, it is desirable to maintain light intensity at a constant level and to have the spectrum stimulate all photosensitive reactions in plants. It is not generally critical to reproduce natural sunlight as long as plants respond with normal development.

The most easily controlled variable of light is duration (photoperiod). Generally, lights are automatically and abruptly turned on or off with a timeclock. Stepwise changes in light intensity are achieved in some chambers by turning on or off groups of lights in sequence, to simulate dawn and dusk. Since these are common human experiences one would intuitively think there was some important plant response associated with this condition. An examination of the literature did not reveal to this author any necessity for stepwise changes.

The control of flux density and spectrum is more difficult. Most plant chambers create light by combining fluorescent and incandescent lamps. Using these two lamp types, various spectra can be developed and a range of intensity can be achieved. The properties and limitations of these and other light sources must be considered in developing a satisfactory light source for any particular need.

Incandescent Lamps

Incandescence is the light created by a heated body. Therefore incandescent light is termed black body radiation and the spectrum depends on the temperature and exposed surface of the heated element. Most incandescent lamps use a tungsten filament. The electrical resistance of tungsten increases with increasing temperature; thus, if the current through the filament remains constant, the temperature increases rapidly to incandescent temperature and light is emitted. As the temperature of the filament is increased, intensity increases and the spectrum shifts to include shorter wavelengths of light (Fig. 1.5). Because of this shift in spectrum, the amount of light increases faster than the total radiated energy. In other words, the luminous efficacy increases. At the melting point of tungsten ($3655°K$) the maximum luminous efficacy is 53 lumens per watt. If filaments are operated close to the melting point, however, their life is very short and, therefore, most tungsten filament lamps operate at cooler temperatures. The range for gas-filled lamps is from $2700°$ to $3050°K$. It is also evident from Figure 1.5 that most of the energy from incandescent lamps is emitted as infrared radiation.

In plant growth chambers the heat created by infrared radiation must be dissipated and is generally useless to plants. Changing the voltage to an incandescent lamp changes the filament temperature. This technique can be used to regulate light intensity, but it must be noted that spectral shifts also occur (the lower the voltage, the greater the proportion of infrared to visible radiation).

The spectrum of incandescent lamps can be altered by the use of filters, and some bulbs are manufactured with filters. For instance, "Daylight" lamps have bluish glass bulbs, which absorb some of the red and yellow, and they emit a light more representative of natural daylight. However, the intensity of total light emitted by these altered lamps is always significantly reduced. Other colored lamps may find specialized use in plant growth chambers, but generally only white or clear bulbs are used. Inside frosting is applied to incandescent lamps to diffuse the light in an effort to eliminate hard shadows. There are two opposing viewpoints regarding frosted lamps among growth chamber users. One holds that where incandescent lamps are used to supplement fluorescent lighting, clear lamps should be used so that the lamp area is totally transparent to the fluorescent lamps behind them. Others argue that clear lamps act as point sources of light and tend to

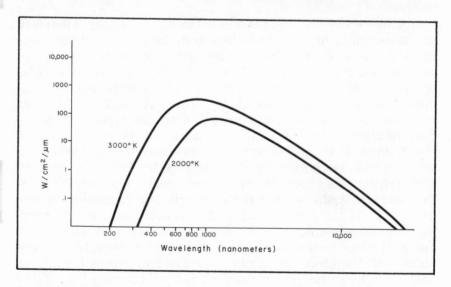

Fig. 1.5. Spectrum of black body radiation at temperatures typical of incandescent lamp filaments

cause more serious irregularities of light intensity in the chamber. Frosting has very little effect on light output from a lamp and, in the author's judgment, the benefit gained by partially diffusing the light outweighs the disadvantage of getting in the way of fluorescent lights. These differences are certainly not serious and make little difference in the resultant light in a chamber.

A relatively new development in incandescent lighting is the halogen-filled lamp. Although the concept has been understood for many years, no practical method was developed until it was found that small quartz envelopes provided the proper conditions. As in all incandescent lamps, tungsten is boiled off the filament. In regular lamps it is deposited on the inside of the lamp and appears as blackening or darkening. In halide lamps the tungsten combines with the halide to form a gas. When this gas gets near the very hot tungsten filament, it is decomposed and the tungsten is redeposited on the filament. Thus the filament is regenerated and theoretically should last forever. Unfortunately, tungsten redeposition is not uniform and parts of the filament eventually become thin and break. The average life of these lamps is from 2000 to 4000 hours, compared to 700 to 1000 hours for most incandescent lamps. Halide lamps are not more efficient—about 22 lumens per watt—but because lamp blackening does not occur, light output remains higher throughout the lamp's life.

The addition of a parabolic aluminized reflector (PAR) to either a regular incandescent lamp or a halogen cycle lamp has some specific advantages in plant lighting. The reflector allows infrared to penetrate and thus be lost to the rear while visible light is reflected toward the front. These lamps are also manufactured with dichroic filters that transmit white light, or they can have special lenses to give blue, green, or red spectra. PAR lamps do not eliminate infrared radiation, but do reduce it sufficiently to make them very attractive in special studies.

Incandescent lamps typically show a decrease in radiant output with time of operation. Figure 1.6 shows that at the end of the expected life, the normal lamp puts out only 85 percent of its original light.

Some lamps are advertised as longer service or long-life bulbs. Longer life is obtained by operating the lamp filament at a lower temperature than normal, or by operating 130 volt lamps on a 110 or 120 volt line. This method shifts the spectrum, decreases light intensity, causes a lower luminous efficacy, and is generally false

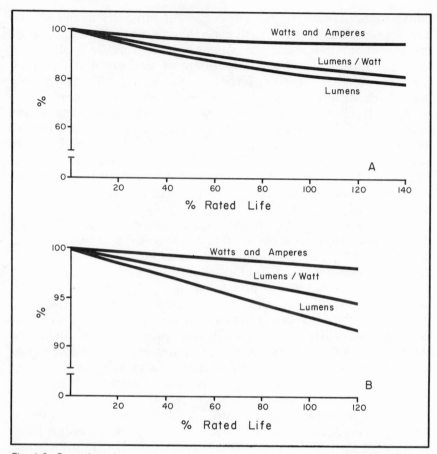

Fig. 1.6. Operating characteristics of lamps as a function of burning time. (A) regular incandescent lamps. (B) tungsten-halogen lamps. (*IES Lighting Handbook* 1972.)

economy. These lamps are designed specifically for situations where changing bulbs is very difficult or expensive and they have no particular value in growth chambers.

Fluorescent Lamps

Fluorescent lamps have been the primary light source in growth chambers for several reasons. When placed close together, they form a continuous, generally uniformly distributed light source. The output of photosynthetically active radiation is high and the spectrum generally matches the requirements of plants. Another important reason is that fluorescent lamps are most efficiently operated at about 32°C and very little of their output is in the infrared range. They produce light with a minimum amount of plant heating. Fluorescent lamps are generally constructed from long glass tubes containing mercury vapor at low pressure with a small amount of an inert gas (generally argon). When the electrodes at each end are supplied the proper voltage, an electric arc is produced through the mercury vapor. Mercury is activated in the arc, and when it drops back to the ground state, energy is radiated primarily in the ultraviolet (253.7 nm), although other wavelengths are present including some visible. The inner walls of the tube are coated with fluorescent powders (phosphors), which are activated by the ultraviolet radiation and fluoresce in the visible region. By blending different phosphors, various spectra can be achieved. Many phosphors are available but not all respond to the primary energy emitted by the mercury arc. The phosphors most commonly used are shown in Table 1.1. The spectral energy distribution (SED) for various lamps is shown in Figure 1.7. Deluxe cool white and deluxe warm white lamps have a broader spectrum than cool white by the addition of red components, although about 30 percent of the light output (compared to standard cool white lamps) is sacrificed. Cool white lamps have also been shown to be the most efficient of the available fluorescent lamps for dry matter production when com-

Table 1.1. Fluorescent chemicals

Phosphor	Lamp color	Exciting range*	Sensitivity peak*	Emitted range*	Emitted peak*
Barium silicate	Black light	180–280	200–240	310–400	346
Barium strontium— magnesium silicate	Black light	180–280	200–250	310–450	360
Cadmium borate	Pink	220–360	250	520–750	615
Calcium halophosphate	White	180–320	250	350–750	580
Calcium silicate	Orange	220–300	253.7	500–720	610
Calcium tungstate	Blue	220–300	272	310–700	440
Magnesium tungstate	Blue-white	220–320	285	360–720	480
Strontium halophosphate	Greenish-blue	180–300	230	400–700	500
Strontium orthophosphate	Orange	180–320	210	450–750	610
Zinc silicate	Green	220–296	253.7	460–640	525

Source: General Electric 1970.
*nanometers.

parisons were made in controlled studies (Biran and Kofranek 1976). Most plant growth chambers employ cool white but some use daylight, warm white, or sign white depending on particular needs of the experiment or desires of the experimenter. Many chambers use fluorescent lamps supplemented with incandescent lamps to provide radiation primarily in the red and infrared region. Figure 1.8 shows a spectrum obtained in a growth chamber by using a combination of 33 cool white, 1500 milliampere (ma) fluorescent lamps (3300 watts) plus six 200 watt (1200 watts) incandescent lamps. At 70 cm from the light cap the radiant flux density between 400 and 700 nm was 13.1 mW · cm^{-2} with both incandescent and fluorescent and 9.8 mW · cm^{-2} with fluorescent lamps alone. The corresponding PPFD and photo-

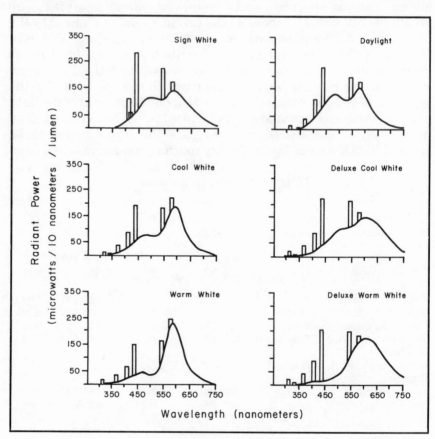

Fig. 1.7. Spectrum of representative fluorescent lamps. (General Electric 1970.)

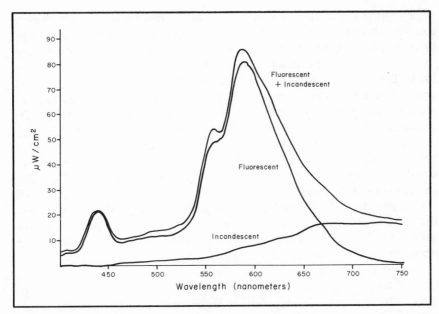

Fig. 1.8. Spectrum of cool white fluorescent lamps alone and supplemented with incandescent lamps, measured with an ISCO scanning spectroradiometer

metric measurements were 35.0 nE cm^{-2} s^{-1} and 24200 Lux respectively.

Fluorescent lamps are generally made with three different types of electrical loadings: 400 ma; 800 ma (also referred to as high output lamps [HO]; and 1500 ma (also referred to as very high output [VHO] or super high output [SHO]). In most growth chambers, 1500 ma lamps are used. Although 800 ma lamps are adequate for germinating many plants, they generally do not yield satisfactory growth. The 1500 ma lamps are rated at approximately 30 watts per foot of tube length; a four-foot lamp requires approximately 120 watts, but an eight-foot lamp requires only 215 watts.

An important part of all fluorescent lamps is the ballast. It provides sufficient voltage to start and operate the lamp. After the lamp is started the gas in the tube ionizes and becomes an electrical conductor. Resistance, therefore, rapidly decreases and unless regulated, the current would increase and burn out the lamp. The ballast controls the current by using a choking coil. Ballasts are a major source of heat and are generally mounted externally and ventilated separately from the lamps in growth chambers.

Fluorescent lamp life is determined by the emissive coating on the electrodes. During normal operation this material is evaporated from the filament and during start-up extra amounts are eroded. Thus the number of starts and length of operation affect lamp life. Lamp life is calculated on the basis of three hours of operation per start. The number of starts affects lamp life as shown in Figure 1.9. Lamp life is defined as that period when 50 percent of a representative sample of lamps will burn out.

Light output decreases with time operated. The output decreases rapidly during the first 100 hours and then more slowly until the lamp burns out. Figure 1.10 shows the decrease in light production with time operated. After 6000 hours (1 year at 16 hours per day) the lamp generally puts out only 70 percent of its original intensity. Lamp life is also influenced by operating line voltage. Voltages either higher or lower than those specified shorten lamp life and reduce efficiency.

Plant growth lamps. Some fluorescent lamps contain special phosphors designed to match the spectrum of chlorophyll absorption. The concept is to produce light that plants will use most efficiently. Some studies with Gro-Lux ® and other similar lamps suggest that better growth was obtained with these lamps than with warm white lamps, but the majority of studies have not

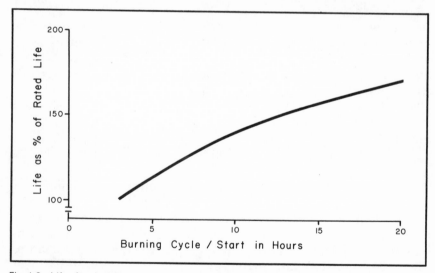

Fig. 1.9. Life of typical fluorescent lamps as a function of burning cycle. Variations from this curve can be expected with lamp loading. (*IES Lighting Handbook* 1972.)

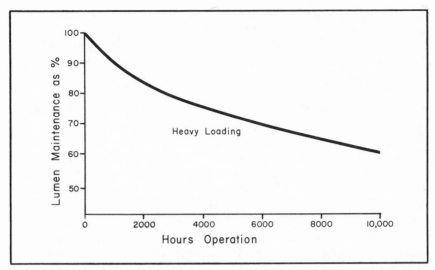

Fig. 1.10. Lumen maintenance curves for typical 1500 ma fluorescent lamps as a function of hours operation. (*IES Lighting Handbook* 1972.)

demonstrated any advantage to these lamps. The efficiency of photosynthesis in terms of incident radiation produced by plant growth lamps may be the same as or higher than for other lamps, but the efficiency based on input electricity is lower. Plant growth lamps have a lower light production efficiency and in general produce an optimum spectrum by the depletion of some bands, not by the addition of others. The efficiency of photosynthetically active light from plant growth lamps is less than for cool white fluorescent lamps. Thus, although plant growth lamps can be used successfully to produce healthy plants, more lamps must be used to obtain similar rates of growth. Also, the red phosphor (magnesium fluorogermanate) used in these lamps makes them more expensive to manufacture than white lamps, thus the advantages are primarily aesthetic; their disadvantage is their greater cost and their lower total light output (as measured by PPFD).

High Intensity Discharge Lamps (HID)

High intensity discharge lamps are discharge devices that are well stabilized, including mercury, metal halide, and high pressure sodium lamps. High intensity discharge lamps produce intense light that may not be obtainable in other ways, and use of these lamps in growth chamber is increasing. They make possible higher intensity irradiation than fluorescent or incandescent lamps. The lamps are discrete point sources, and it is more difficult to obtain uniform light distribution over a growing area than with fluorescent lamps. The predominance of radiation in discrete line spectra have encouraged researchers to combine two different types of HID lamps in growth chambers to obtain a more balanced spectrum. For plant species or particular research studies requiring intensities of radiation above 50 nE cm^{-2} s^{-1}, these lamps are recommended even though it may be difficult to obtain uniform spectral intensities over the plant growing area.

Mercury lamps. Light from mercury lamps is produced by arcing electricity through mercury vapor much as in low pressure fluorescent lamps. They differ from fluorescent lamps in that the arc length is shorter, temperature is higher, and the vapor pressure of mercury is greater. With increased vapor pressure, the visible emission lines are intensified and the ultraviolet lines diminished. The quartz arc tube transmits all wavelengths produced but the outer glass bulb cuts out most wavelengths below 300 nm. The clear mercury lamp produces a bluish-white light with very little red. A typical spectrum is shown in Figure 1.11A. Phosphors on the inner surface of some types of mercury lamps are used to alter the spectrum as seen in Figure 1.11B. These are termed color improved or mercury phosphor lamps.

Most mercury lamps have an average life of 24,000 hours, but as with other electric light sources, the output decreases with hours operated. After 12,000 hours, the output is from 70 to 85 percent of the original, depending on the particular lamp. The PAR efficiency as indicated in Table 1.3 is about the same as plant growth lamps but significantly lower than HID metal halide lamps and white fluorescent lamps. Unlike the output of fluorescent lamps, mercury lamp output is not significantly affected by ambient temperature. The outer bulb acts as a temperature shield and the temperature of the quartz arc tube remains nearly constant.

Metal halide lamps. Metal halide lamps are similar to mercury lamps except the arc tube contains various metal halides in addition to mercury. Halides are usually iodides of thorium, thallium, or sodium. During operation the metalic halides are vaporized and produce the characteristic line spectra.

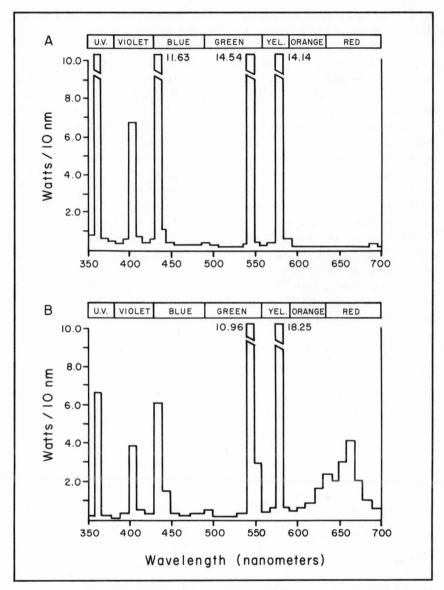

Fig. 1.11. Spectral energy distribution of (A) 400-watt clear mercury lamp and (B) 400-watt color improved mercury lamp (H33-1GL/C). (Sylvania 1972.)

The metal halide lamps appear to be the most useful type of HID lamp for plant growth. The emission spectra is almost continuous over the 400 to 700 nm wavebands. Plant response data indicate that the combination of sodium HID lamps with metal halide lamps provides superior plant growth to the use of metal halide lamps alone. In addition, the PAR efficiency of metal halide HID lamps is less than for sodium HID lamps. The use of metal halide lamps in careful research studies also is discouraged because there are significant spectral distribution shifts with aging that would produce variable plant growth responses.

High pressure sodium lamps. A high pressure sodium lamp is of particular usefulness for plant growth because of its high PAR efficiency, as shown in Table 1.2. It is superior to all incandescent, fluorescent, and other HID lamps. On the other hand, emission from the lamp is completely devoid of radiation in the 400–500 nm waveband—radiation required for certain morphogenic responses of plants. Thus, it is necessary to use either mercury or metal halide HID lamps in conjunction with sodium lamps in growth chambers and growing rooms. Most commonly they are used with mercury or improved mercury lamps.

Table 1.2. **Photosynthetic radiation efficiency of lamps**

Lamp type	Input power (Watts)	Efficiency (400–700 nm) (mW/W)
Incandescent	25	39
	40	45
	60	57
	100	69
	200	79
Fluorescent		
Cool white	46	204
Cool white	225	204
Warm white	46	199
Plant growth A	46	127
Plant growth B	46	146
High intensity discharge		
Clear mercury	440	124
Mercury phosphor	440	131
Metal halide A	460	204
Metal halide B	460	225
Sodium	470	245

Source: Campbell, Thimijan and Cathey 1975.

Other Lamps

Low pressure sodium lamps.   The usefulness of low pressure sodium lamps for supplemental lighting is being demonstrated in greenhouses, where a small point source lamp is advantageous. However, the intensities are not great enough for them to be as useful as fluorescent or HID lamps in growth chambers.

Self-ballasted mercury lamps.   Most major lamp companies supply self-ballasted mercury lamps and they can be utilized in an a-c line without the need of a ballast. The lamps provide radiation from both a tungsten filament and from mercury arc emission. These lamps cannot be recommended in growth chambers over fluorescent or HID lamps because they emit a large quantity of radiation of wavelengths longer than 700 mm. Long wave radiation is not photosynthetically active and thus is an additional heat load to the chamber.

Xenon lamps.   Xenon lamps have the potential of providing irradiation for plants that most nearly duplicates the spectrum of sunlight and most nearly approximates the intensity of sunlight. Unfortunately, they are expensive and generate large quantities of long wave radiation that increases the cooling requirements in the chamber. Thus, they have had only limited use in a few laboratories.

References

Bernier, C. J. 1962. Measurement techniques for the radiant energy requirements of growing plants. Sylvania Electric Products Inc. Reprint 48.

Bickford, E. D., and S. Dunn. 1972. *Lighting for Plant Growth.* Kent State University Press, Kent, Ohio. 221 pp.

Biran, I., and A. M. Kofranek. 1976. Evaluation of fluorescent lamps as an energy source for plant growth. *J. Amer. Soc. Hort. Sci.* 101(6):625–628.

Campbell, L. E.; R. W. Thimijan; and H. M. Cathey. 1975. Spectral radiant power of lamps used in horticulture. *Trans. ASAE* 18:952–956.

Committee on Plant Irradiation of the Nederlandse Stichting voor Verlichtingskunde. 1953. Specification of radiant flux and density in irradiation of plants with artificial light. *J. Hort. Sci.* 28(3): 177–184.

Davis, N., and A. W. Dimock. 1961. Preliminary findings of Bioclimatic Laboratory development project at Cornell University. Environmental Growth Chambers, Chagrin Falls, Ohio.

Federer, C. A., and C. B. Tanner. 1966. Sensors for measuring light available for photosynthesis. *Ecology* 47:654–657.

Gaastra, P. 1959. Photosynthesis of crop plants as influenced by light, carbon dioxide, temperature, and stomatal diffusion resistance. *Mededelingen van de Landbouwhoogeschool te Wageningen, Nederland* 59(13):1–68.

General Electric. 1970. Fluorescent lamps. TP-111.

Hammer, P. A., and R. W. Langhans. 1972. Experimental design considerations for growth chamber studies. *HortScience* 7:481–483.

Hollaender, A. 1956. *Radiation Biology*. Vol. III. McGraw-Hill, New York.

IES Lighting Handbook. 1966. Illuminating Engineering Society, New York.

———. 1972. Illuminating Engineering Society, New York.

Klein, R. 1973. Determining radiant energy in different wavelengths present in white light. *HortScience* 8:210–211.

McCree, K. J. 1966. A solarimeter for measuring photosynthetically active radiation. *Agric. Meteor.* 3:353–366.

Norris, K. E. 1968. Evaluation of visible radiation for plant growth. *Ann. Rev. Plant Phys.* 19:490.

Shibles, R. 1976. Committee Report. Terminology pertaining to photosynthesis. *Crop Sci.* 16:437–439.

Sylvania. 1972. Mercury lamps. Eng. Bul. 0–346.

Tibbitts, T. W.; J. C. McFarlane; D. T. Krizek; W. L. Berry; P. A. Hammer; R. W. Langhans; R. A. Larson; and D. P. Ormrod. 1976. Radiation environment of growth chambers. *J. Amer. Soc. Hort. Sci.* 101(2):164–170.

Tibbitts, T. W. 1977. Personal communication based on unpublished data.

Tyler, J. E. 1973. Lux vs. quanta. *Liminol. Oceanogr.* 18(5):810.

van der Veen, R., and G. Meijer. 1959. *Light and Plant Growth*. Centrex, Eindhoven, Holland. 159 pp.

Chapter 2   DOUGLAS P. ORMROD

Temperature

Growth chambers provide a convenient method for studying the effects of temperature, separately or in combination with other environmental factors (Hudson 1957, Went 1957), and for establishing optimum temperature conditions for plant growth. The study of temperature effects on plants is much less complicated in controlled environments than in the field where temperatures vary from hour to hour and day to day. In a growth chamber, temperature regimes can be programmed in a definite way and the effects can be easily observed.

Temperature may be defined as the energy state of an object. Differences in temperature between an object and its surroundings will determine whether energy in the form of heat will flow into or out of that object (Platt and Griffiths 1964).

Heat Transfer

Heat energy can be transmitted as radiation, by moving air (convection), or from molecule to molecule (conduction). The temperature of an object can also change as a result of condensation or evaporation of water at the surface, which releases or requires heat. There are many methods available for the measurement of temperature and heat transfer or flux (Doebelin 1966).

Radiation is the direct transfer of heat by infrared radiation between two objects with different temperatures. The amount of radiation is defined by the equation:

$$q = F_E \cdot F_A \cdot A \, (T_1{}^4 - T_2{}^4)$$

where $\acute{F}_E$ = emissivity factor of the two materials,
F_A = relationship of the two objects,
A = cross section area of the radiating objects in 'view' of one another,
T_1 and T_2 = absolute temperatures of the two objects.

Perhaps the two features of greatest interest are the importance of the cross section exposed and of the fourth power of the absolute temperatures.

Convection is the transfer of heat using moving air or liquid as the medium of exchange. Convection is defined by the formula:

$$q = hc \cdot A \cdot \delta t$$

where hc = film coefficient,
A = surface area of the object,
δt = temperature difference between the object and the surrounding air or liquid.

Whenever there is a temperature difference between a plant and the surrounding air, convection will occur with movement of air past the plant. With slight temperature differences the flow of air past leaves is laminar, but as temperature differences increase the flow becomes turbulent. The position of the leaf and of the heated surface is also important, as shown by the following formulas.

Natural convection over vertical surfaces:
Laminar: $hc = 0.19\,\delta t^{1/3}$
Turbulent: $hc = 0.29\,(\delta t/L)^{1/4}$
where L = length of surface.

Natural convection over horizontal surfaces:
Heated surface up:
Laminar: $hc = 0.22\,\delta t^{1/3}$
Turbulent: $hc = 0.27\,(\delta t/L)^{1/4}$
Heated surface down:
Laminar: $hc = 0.12\,(\delta t/L)^{1/4}$.

Convection heat transfer can be speeded up markedly by rapid or forced air movement, provided outdoors by the wind, indoors by fans.

The following is a typical formula for film coefficients with forced air, in this case for horizontal cylinders:

$$hc = 0.24 \, (Kf/L) \, (LVP/\mu)$$

where Kf = thermal conductivity of fluid (air in the case of
terrestrial plants),
L = diameter of cylinder,
V = velocity of fluid (air) stream,
P = fluid (air) density,
μ = fluid (air) viscosity.

The most interesting features of convection are its dependence on
surface area (convection essentially envelops the plant) and the
great improvement in heat transfer obtained by using forced air
movement. The upward movements of warm air by natural convection (during the light period when leaves may be much
warmer than the surrounding air) are often referred to as plumes
or eddies. During the dark, the convective flow of heat is to the
leaves.

Conduction is the movement of energy from molecule to molecule. It is a slow process and is generally of much less importance
to plant heating and cooling than radiation and convection. Conduction is defined by:

$$q = KA \, \delta t$$

where K = coefficient of conductivity,
A = area of heat flow,
δt = temperature difference.

Effects of Temperature on Plant Growth and Development

While many effects of temperature on plants are well known,
there remain many incompletely known or unknown temperature
responses (Evans 1963). Temperature controls overall plant
growth and development; it controls the growth of individual cells
and development of complex organs; and it exerts a controlling
effect on every enzyme-mediated pathway of metabolism (ASH-
RAE 1972). The rate of growth and the form that a plant takes are
a function of the effects of temperature on the separate processes
of photosynthesis, transpiration, respiration, mineral and water
absorption, and floral initiation. Temperature exerts a controlling
influence at every stage of the plant's life cycle from seed germination to fruit ripening. Downs and Hellmers (1975) have reviewed many aspects of temperature control of plant growth and
development.

Temperature interacts with other environmental factors to influence growth and development. For example, photoperiodic effects may be enhanced or eliminated by the temperature in the light and dark periods; low light intensity may reduce the optimum temperature for growth; and the response to raised carbon dioxide levels may be enhanced by increased temperature. Higher temperatures commonly increase the moisture stress within the plant.

Plant cells maintain normal metabolic activity only within a certain temperature range, which can vary with species, cultivar, and growth stage, as well as with the previous temperature history of the plant. The best or optimum temperature for germination and growth of seedlings is frequently higher than that required for best growth at a later stage. Even the processes of flower initiation and fruit formation may have different optimum temperatures.

Many plants have been found to grow and develop better when there is a daily fluctuation in temperature. Such plants are said to have a thermoperiodic requirement and the phenomenon is called thermoperiodism. In recent years, evidence has been accumulated that, for certain of these plants, the thermoperiodic requirement is negated if plant moisture stress is minimized during the light period.

Plants are killed by extreme temperatures (Levitt 1972). Both the temperature and duration of exposure are important determinants of lethality, especially at high temperatures. Excessively high temperatures destroy proteins, inactivate enzymes, and disintegrate cell membranes. At temperatures below freezing, ice crystal formation may cause direct physical injury to living cells. Cold injury may also result from the desiccation that occurs when water is withdrawn from cytoplasm in the formation of ice crystals between the cells. The tolerance of living cells to extremes of low or high temperatures may be increased by acclimation, or progressive slow change of temperature over time.

Soil temperature affects plant growth and development and may have effects independent of air temperature. The ability of a plant to absorb water and nutrients from the rooting medium is highly dependent upon soil temperature.

Plant Temperature

The above-ground plant temperature is governed by the balance of incoming and outgoing radiation, the evaporative cooling effect of transpiration, and the conductive-convective effect of air movement past the plant. The principal radiation sources in growth chambers are the lamps. Considerable refrigeration via forced convection (fan-forced air movement) is required to remove the energy contributed to plants by the incoming radiation from the lamps. Under conditions favouring rapid transpiration (rapid air movement and moderate or low humidity), the evaporative cooling effect may make the plants cooler than the surrounding air. Thus, the plant temperature/air temperature differential may be as much as ±2°C, even with forced convection. Leaf temperatures may be lower than stem and fruit temperatures, because of the cooling effect of transpiration on the leaf surface that does not occur on stems and fruits. Plant temperatures can be estimated by fine-wire thermocouples either inserted in the tissue or closely appressed to the plant surface in a position shaded from the direct radiation of the lamps (Long 1968).

Root temperature may differ from shoot temperature because water evaporates from the surface of the rooting medium and because the air and soil do not maintain a temperature equilibrium during diurnal air temperature cycling. Rooting medium temperature changes more slowly than air temperature. The nature of the container also affects root temperature. A porous container will be cooler than a nonporous one because of evaporation from the sides. This evaporation may be more significant in growth chambers than in greenhouses because of the much greater air movement in a growth chamber. Container color will also have some effect because dark-colored pots will absorb more radiant energy. The temperature of the nutrient solution will affect root temperature immediately after nutrient feeding.

Temperature Measurement

Many devices are available for temperature measurement, including gas, liquid-in-glass, liquid-in-metal, and deformation thermometers, all of which depend on coefficients of heat expansion. Mercury-in-glass is the chief instrument used in the laboratory. Its fragility, large size, and slow response time are disadvantageous in growth chamber work. Deformation thermometers based on bimetallic strips and Bourdon tubes based on curvature of a mercury-filled tube are also large and have slow response times.

Electrical thermometers are preferred for use in growth chamber research. Thermocouples operate on the principle that current flows through wires of two dissimilar metals joined together to form a circuit if the two junctions are at different temperatures. The electromotive force (emf) established in the circuit depends upon the nature of the two metals and the difference in temperature between the two junctions. One junction is kept at a standard temperature (usually 0°C) and a sensitive galvanometer is placed in the circuit to measure the emf in millivolts. The system can be calibrated to relate voltage to temperature. Thermocouples are widely used to measure temperature and can also be used to measure radiation, humidity, and air flow. The thermocouple has great sensitivity to temperature, it can be made as small as desired within the limits of wire size, and the meter can be located at a distance from the point of observation.

Resistance thermometers operate on the principle that the electrical resistances of metals and metallic oxides change with temperature. Thermistors are a type of resistance thermometer in which metallic oxides with platinum alloy wire leads are arranged in a circuit. The resistance varies markedly with temperature and can be measured electronically. Thermistors are available as beads, rods, discs, and probes. Thermistors have greater sensitivity than thermocouples and they are monitored with more rugged and less expensive external indicating equipment.

Temperature Control Equipment

The growth chamber facility must include effective systems for control and monitoring of temperature. Temperatures must be uniform throughout the chamber and the control system must make adjustments automatically according to a predetermined temperature regime. The facility must have direct cooling and heating with separate cooling coils and heating elements located in the air ducts or a single heat exchanger used for both cooling and heating (Schultze 1972).

The temperature range available in growth chambers is dependent on the size of the refrigeration and heating systems. Temperatures available in standard commercial growth chambers are about 10° to 35°C with lights on and 5° to 30°C with lights off. Units operated at lower temperatures may have serious problems with frost build-up on coils and will thus require dual coolers with defrost cycling.

Humidity levels in growth chambers will be greatly influenced by temperature as well as by the nature of the temperature control equipment (see Chapter 3, Humidity). The design and operation of the temperature control equipment may likewise be influenced by the desired level of humidity and precision of control (Schultze 1972). For an extensive review of the nature of temperature control systems and technical problems in their incorporation into growth chambers, the reader is referred to Downs 1975.

Cooling Systems

Growth chambers may be supplied with self-contained compressor systems for cooling or they may be cooled by a liquid such as chilled water or ethylene glycol (antifreeze) circulated from a central refrigeration system or modulated control system. The on-off control system is less expensive but there may be considerable fluctuation in temperature and humidity between cycles. The modulated control system is more desirable for growth chambers because there will be little temperature and humidity fluctuation. The temperature approaches a modulated control point and remains steady.

A central refrigeration system or secondary cooling system may be used if there is a considerable number of growth chambers to be served. The one or two large compressors needed are more efficient than many small compressors. The chilled water or antifreeze solution that is circulated throughout the growth chamber area enters the cooling coils in each chamber according to a modulated valve system in order to maintain constant chamber temperatures.

The cooling coils are usually placed in the main air duct, away from the plants. A large volume of air must be moved past these coils to allow temperature control in the chambers. Chilled water is useful for the maintenance of temperature down to about 10°C but ethylene glycol or built-in compressor units must be used for lower temperatures. Chambers with air-cooled refrigeration units must have sufficient air circulation around the external heat exchange coils to ensure proper removal of heat, and care must be taken to see that the room in which they are located is well ventilated. Water-cooled compressors have particular water requirements according to the size of the compressor unit and provision must be made for handling the cooling water. Many institutions have restrictions on the use of water for cooling, and external evaporators may be required, thus allowing the cooling water to be recirculated. Health safety codes often require that the discharge water from the heat exchanger be emptied into an open drain.

Heating Systems

The heating system must provide sufficient heat to maintain the desired temperatures in the plant growing area. During the light period, however, virtually all of the electrical energy used by the lamps appears in the growth chamber in the form of heat. The problems of temperature maintenance during the light period are largely associated with providing adequate refrigeration, unless high chamber temperatures are desired. The heating system will be required normally during dark periods and then only if the desired temperature is above the temperature of the room in which the chambers are located.

Electrical resistance heaters in multiples of 1-kilowatt units are normally used. They are usually switched on or off upon demand by the thermostats in the chambers. However, they may be equipped with motorized variacs that control air temperatures by varying the voltage across the heaters. The heaters are usually located in the main air conditioning ducts in a position where direct radiation to the plants will not occur.

A single heat exchanger may be used in the air conditioning duct for both heating and cooling. The temperature of the liquid in the heat exchange coil is externally controlled. The use of such a temperature conditioning circuit has less effect on humidity because the coil is at a constant temperature rather than alternately cold and warm, thus resulting in little or no condensation.

Thermostats

A temperature control system requires the use of thermostats to sense changes in temperature and to start or stop the heating and cooling systems. The precision of control depends on the type of thermostat, on the means of connecting the thermostats and the heaters and coolers, and on the types of heaters and coolers.

Ideally, the control system should react immediately to temperature change in the chamber, but the thermostat sensor always takes some time to react to a temperature change. There is also an additional delay while the heating or cooling equipment responds to the sensor. The "on-off" control systems result in the greatest time lags and largest temperature oscillations. Oscillation can be reduced with more complex control systems. The best systems provide a smooth modulating or proportioning action in response to fluctuations in air temperature. A feedback system is used to minimize time lags.

Growth chambers may be fitted with several types of thermostats. Selection criteria are sensitivity, durability, and price. Electronic controls provide the most precise regulation, while less expensive mechanical thermostats cannot maintain as close a temperature differential.

Electronic temperature sensors with solid-state transistorized circuits are available with a differential as low as ±0.1°C. The sensors are generally thermistors for which the electrical resistance changes with change in temperature. Fine wire resistance thermostats give precise control but problems with damage to the sensor wires have curtailed their use. Direct digital control of the temperature program may be introduced where electronic sensors are used (Trickett and Moss 1972).

The two types of mechanical thermostats normally used are based on the expansion and contraction of liquids and metals with changes in temperature. Liquid expansion mechanical thermostats are based on the principles that liquids have a large coefficient of expansion and their expansion rate is uniform at all temperatures. The differential is thus the same at all points over the temperature range. As the air temperature varies, the liquid in a sensing bulb expands or contracts causing a bellows to activate an electrical switching mechanism or a pneumatic system. In the metal expansion thermostats, strips of two different metals are joined together so that their different rates of expansion and contraction cause the strip to bend and to activate an electrical switching mechanism or a pneumatic system. Mechanical thermostats have long lag periods and the amplitude of air temperature oscillations can be substantial. They are not very satisfactory for precisely controlled growth chamber research.

Temperature Programmers

Temperature regimes required for different experiments may vary from one uniform temperature day and night to continuously changing temperatures over a 24-hour period. Continuous constant temperatures (within the level of control in the system) can be maintained by a single thermostat, but this system would not be satisfactory if a day/night temperature fluctuation is to be imposed. Day/night temperature cycling is most commonly provided by two thermostats. A 24-hour clock automatically switches the chamber temperature control from one thermostat to the other at the desired time.

Continuous changes in temperature can be obtained by using a cam-operated control system. Templates are cut for the specific temperature regime desired over a 24-hour period. The thermostat cam arm follows the edge of the template to control the desired temperature over time. A camless temperature programmer consists of a set of electronic or mechanical thermostats and a timeclock. The timeclock selects the particular thermostat, which is set to give the desired temperature for that time period. This system does not allow for a gradual change from one temperature to the next, as the timeclock will switch to the next thermostat at the desired time in a steplike fashion.

Safety Controls

Growth chambers should be equipped with safety controls that limit the lower and upper range of temperature in the chamber. In the event of malfunction of the refrigeration or heating systems or the control systems, the safety control will turn off all equipment and should sound an alarm or activate a warning light when the temperature reaches the set limits. Such a limit system is necessary to prevent damage to the chamber as well as to prevent destruction of plant material. The temperature limit controls should be set just above and below the temperature range of the experiment. If the safety control is activated, the chamber temperature will approach the ambient temperature until the problem is corrected. If the malfunction is of short duration plant materials could still be used in some cases.

Design Considerations

Location of temperature sensors. Sensors for temperature control should be inside a box or tube ventilated by a fan. The box or tube should be located to obtain a representative air sample from the chamber and the sensor should be shielded from direct or reflected radiation from the lamps, as radiation might raise the sensor temperature above the air temperature during the light period. The rate of airflow over the sensor should be great enough to ensure that air temperatures at the sensor follow air temperatures over the plants. For studies in which the temperature of the plants is of particular importance, a shielded thermocouple should be located at plant level and used to calibrate the temperature control sensor to the desired level.

Uniformity of temperature and air circulation.   Ventilation and air distribution patterns are established by fans and modified by the form of the growth chamber and its contents. These patterns, as well as the temperature control system, affect the horizontal and vertical temperature uniformity. The degree of temperature uniformity may be tested with a shielded thermocouple or thermistor probe using grid patterns. Both spatial variation and the variation between cooling and heating cycles should be determined.

Air intake.   Most growth chambers have an adjustable air intake port. Many chambers also have leaks through which outside air can enter. The incoming ambient air may have an effect on the temperature relations in the chamber. The difference between exterior and interior temperature will determine the heating and cooling loads. When chamber temperatures are low and the external air temperatures are very high, or vice-versa, the opening of the port should be minimal to prevent an excess load on the refrigeration or heating systems. Some sealing of leaky chambers may also be necessary. Where carbon dioxide is not artificially supplied the amount of incoming air should be just sufficient to replenish carbon dioxide for photosynthesis in the growth chamber.

Operation of Temperature Control Systems

Calibration.   The calibration of temperature controllers should be checked with a test thermometer of the desired precision, preferably a remote sensing thermometer. The sensor of the test thermometer should be placed inside the same housing or the same type of ventilated box or tube as the temperature sensor of the thermostat. The test thermometer should be calibrated electronically or with water and ice.

Temperature monitoring and recording.   The temperature of a chamber should be monitored continually to observe any irregularities during each diurnal cycle. Temperature may be monitored and recorded by devices that have the same operation as thermostats. Mechanical methods, such as liquid or metal expansion, will be slower to respond to temperature changes than methods based on the use of thermistors or thermocouples. As for temperature control systems, the temperature sensors should be ventilated and shielded to provide representative responses and minimize errors arising from radiation.

In addition, the sensors for both control and recording should be calibrated or readjusted using a portable electronic temperature sensor placed beside or at the top of the plant canopy for short periods to ensure that the desired plant temperature is maintained.

References Cited

ASHRAE. 1972. Environmental control for animals and plants: Physiological considerations. Pages 151–166 in *ASHRAE Handbook for Fundamentals*. Ed. C. W. MacPhee. Amer. Soc. Heating, Refrig., and Air Cond. Eng., New York.

Doebelin, E. O. 1966. *Measurement Systems: Application and Design*. McGraw-Hill, New York. 743 pp.

Downs, R. J. 1975. *Controlled Environments for Plant Research*. Columbia University Press, New York. 175 pp.

Downs, R. J., and H. Hellmers. 1975. *Environment and the Experimental Control of Plant Growth*. Academic Press, New York. 145 pp.

Evans, L. T., ed. 1963. *Environmental Control of Plant Growth*. Academic Press, New York. 499 pp.

Hudson, J. P. 1957. *Control of the Plant Environment*. Butterworths, London. 240 pp.

Levitt, J. 1972. *Responses of Plants to Environmental Stresses*. Academic Press, New York. 697 pp.

Long, I. F. 1968. Instruments and techniques for measuring the microclimate of crops. In R. M. Wadsworth, ed., *The Measurement of Environmental Factors in Terrestrial Ecology*. British Ecological Society Symposium vol. 8. Blackwell Scientific Publications, Oxford. 314 pp.

Platt, R. B., and J. F. Griffiths. 1964. *Environmental Measurement and Interpretation*. Reinhold, New York. 235 pp.

Schultze, H. A. 1972. Technical problems of test cabinets and chambers for simulation of environmental conditions. In P. Chouard and N. de Bilderling, eds., *Phytotronique et prospective horticole*. Gauthier-Villars Editeur, Paris. 391 pp.

Trickett, E. S., and G. I. Moss. 1972. Measurement and simulation of diurnal and seasonal fluctuations in climate. In P. Chouard and N. de Bilderling, eds., *Phytotronique et prospective horticole*. Gauthier-Villars Editeur, Paris. 391 pp.

Went, F. W. 1957. *Experimental Control of Plant Growth*. Chronica Botanica, Waltham, Mass. 343 pp.

Chapter **3**   THEODORE W. TIBBITTS

Humidity

The importance of humidity levels in regulating plant growth has been well documented in recent years. Plant scientists now recognize that humidity control in growth chambers must be provided to insure uniformity in plant response from study to study. Precise humidity levels are difficult to obtain and also difficult to maintain over long periods of time. Thus, many chambers and facilities are without adequate humidity control for plant research.

The humidity level in growth chambers is characterized by large variations. If no control is provided, the variation between the light and dark period and over each heating and cooling cycle will be quite large. During the light period, cooling is usually maximum and large quantities of water condense from the chamber air onto the cooling coils, thereby, decreasing the moisture content of the air. Between 15° and 30°C, most chambers will be at about 50 percent relative humidity during the light period unless humidification is provided. During the dark period the chambers will be at or near saturation levels if no significant amount of cooling is required. The excessive moisture during the dark periods may cause condensation on walls and in the electrical systems which can lead to short circuits and interruptions. During a single cycle of heating and cooling (usually 1 to 3 minutes) the fluctuations in relative humidity may be as much as 20 to 40 percent.

Low humidity levels during the light period necessitate maintaining adequate levels of soil water, particularly in chambers without humidity control. Evaporative water losses tend to be great in growth chambers, accentuated by the high rates of air velocity that must be maintained for temperature uniformity and by the dehumidification function of the cooling coils.

Characteristics

Humidity refers to the water present in air as a gas, which is commonly termed water vapor. The maximum quantity of water that can be retained in air at temperatures between 0° and 50° is shown in Figure 3.1 and Table 3.1.

Vapor pressure (e). The quantity of water present in air in the gas phase can be stated as the vapor pressure of the water vapor and is expressed in bars of pressure (International unit), millimeters of mercury, or inches of mercury. Vapor pressure is an expression of the activity of vaporized water molecules. At the maximum vapor pressure, called the saturation vapor pressure (es), the air is saturated with water vapor. The saturation vapor

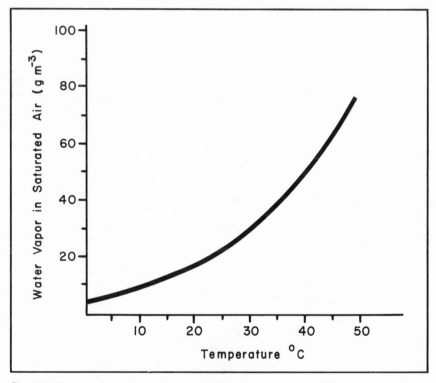

Fig. 3.1. Water content of saturated air at different temperatures (760 mm pressure)

pressure increases with temperatures as shown in **Figure** 3.2 and Table 3.2.

Relative humidity (RH). Relative humidity is the water vapor present in an air mass expressed as a percentage of the saturated vapor content or vapor pressure at a specific temperature and pressure of measurement. Absolute humidity is the mass of water vapor present in a unit volume of air (expressed as grams per liter [g/L], pounds per cubic foot [lb/ft³]). The mass of water vapor per cubic meter of air at different temperatures is given in Table 3.1. Specific humidity is the mass of water vapor present in a unit mass of air (expressed as grams per gram [g/g], pounds per pound [lb/lb]). The absolute humidity and specific humidity are difficult to measure directly, but can be determined from relative humidity or dew point measurements when temperature and pressure are known.

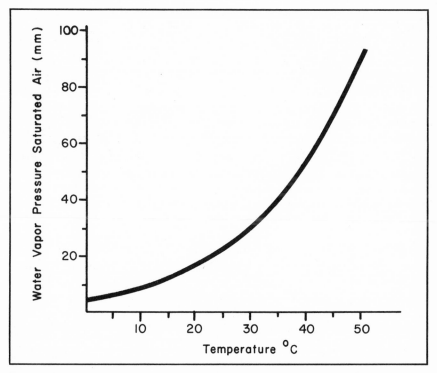

Fig. 3.2. Saturation vapor pressure of water vapor in air at different temperatures (760 mm pressure)

Dew point. The temperature of an air mass at the saturated vapor pressure is the dew point. The conversion of dew-point temperatures to relative humidity levels or vice versa can be determined through use of Tables 3.3, 3.4, 3.5, and 3.6. Dew point is a valuable parameter for the growth chamber scientist for it provides a measurement of the aridity of the air and can give an indication of the moisture stress to which the plants and soil are being subjected.

Saturation vapor pressure deficit. The difference in pressure between the saturation vapor pressure and the actual vapor pressure of the air is expressed as millibars, mm of mercury, or inches of mercury. Table 3.2 provides saturation vapor pressure deficits for air at different relative humidity levels and temperatures between 0° and 50°C.

Influence on Plant Growth

The moisture content of the air has been shown to have a significant controlling function upon the growth and development of plants (Ford and Thorne 1974, Gaastra 1959, Hoffman 1972, Hoffman and Rawlins 1971, Hoffman et al. 1971, Kristoffersen 1963, Krizek et al. 1971, Read 1972, Tibbitts and Bottenberg 1976, Tromp and Oele 1972). The moisture content of the air, or more specifically, the vapor pressure deficit of the air, influences the amount of water transpired from the plant and thus controls significantly the water stress to which the plant is subjected (Hoffman 1972, Kozlowski 1976, Troughton and Slayter 1969). When stomates are open, the rate of diffusion of water vapor from a leaf is proportional to the difference in vapor density between the leaf air space system and the atmosphere (Meidner and Mansfield 1968). Thus, with increasing vapor pressure deficits of the air in the chamber a greater amount of water will be transpired by the plant. The large amount of air movement required for temperature control in chambers maintains maximum diffusion gradients at the surface of the leaves and thus encourages transpiration. Studies with different plants have demonstrated 2 or 3 times as much transpiration at 50 percent RH than at 85 percent RH (Hoffman and Rawlins 1971, Hoffman et al. 1971, Slayter 1973).

The moisture content of the air can alter the temperature of the plant by influencing transpirational cooling of the leaf. The extent of transpirational cooling will vary with the degree of stomatal opening and the rate of air movement passing by the leaves (Gaastra 1959). A 3 to 4°C decrease in leaf temperatures has been demonstrated for plants maintained in atmospheres under large vapor pressure deficits compared to plants maintained under small vapor pressure deficits (Matsui and Eguchi 1972).

The moisture content of the air controls nutrient accumulation in above-ground tissues of plants. Large atmospheric moisture deficits produce large transpiration rates and accumulation of certain nutrients at the tips and margins of leaves that can lead to toxicity and necrosis of the tissues. Continuous atmospheric moisture deficits and rapid leaf transpiration during the entire diurnal cycle can lead to deficiencies of calcium in leaves or tissues that are not transpiring, as in inner head leaves of cabbage (Palzkill et al. 1976).

Levels in Chambers

The humidity gradients across a chamber have been found to be proportional to the temperature gradients in the chamber, but have not been found to be very significant in chamber studies. In areas of the chamber where temperatures are warmer than the average temperature of the chamber, the relative humidity will be the lowest and vapor pressure deficit will be the largest. Where temperatures are cooler, the opposite will be true. Thus, the greater the airflow, the greater the uniformity of airflow across the chamber, the smaller the temperature gradients and similarly the smaller the humidity gradients.

There are several fluctuations of humidity over time that are significant. They occur with heating and cooling cycles, with watering schedules, with light and dark cycles, and with the seasons of the year.

The fluctuations with heating and cooling cycles may be very great, particularly if the cooling system is run at maximum. Measurements in reach-in chambers having electrical conductivity sensors for control have demonstrated 3°C dew-point temperature variations and RH fluctuations of 10 percent within a 90-second period during the heating and cooling cycles. For larger walk-in units with hygroscopic sensor controllers, the fluctuations have been found to exceed 20 percent RH. The humidity decreases with cooling, as moisture condenses on the cool coils, and increases with heating as moisture evaporates from the warming coils. Fluctuations will be less if no humidification systems are operating and the chambers have stabilized near the dew-point temperature of the cooling coils. The variations can be minimized if the pressure on the coolant is reduced as much as possible. Reduction of pressure in the coils maintains a higher coolant temperature and thus the dew-point temperature of the coils will be higher. The variations can be effectively minimized and essentially eliminated through use of proportional control of the pressure on the coolant going into the coils.

An increase in humidity may be realized with watering, and humidity will remain higher as long as there is free moisture on the trays and floor of the chamber area.

The fluctuations over light and dark cycles will be rather large unless humidity control is provided in the chamber. During the dark cycle, much less cooling is required because the heat load from the lighting system has been removed, and thus humidity levels will be increased as less water is condensed out on the cooling coils. Measurements in an empty reach-in chamber without humidification at 20.5°C produced a 9.8°C dew point (51% RH) during the light period and a 16.5°C dew point (78% RH) during the dark period. The humidity level will be higher, particularly during the dark period, when plants are in the chamber. The use of ambient heaters during the dark period will minimize this increase because cooling will be required to remove the heater load. Excess humidification during the dark periods may result in undesirable condensation on control elements and electrical contacts within the chamber. In chambers without barriers, condensation has caused short circuits in the wiring of the lamps and produced chamber failure.

Fluctuations over the course of a year result from the different absolute humidities of ambient air during the summer and winter. The fluctuations will be most noticeable in areas with very cold winters. There is a large leakage of room air into most commercial chambers whether ambient air is being introduced or not. Thus during the winter months the chamber will be drier than during the summer. Because of this fluctuation, humidifiers within a chamber cannot maintain as high a level during the winter as in the summer.

The amount of plant tissue within a chamber can have some effect upon humidity levels. This effect will be greatest under low light intensities and warm temperatures, and during the dark period. High light intensities and cool temperatures require a lot more cooling and the temperature of the cooling coils becomes the determining factor in the control of the moisture level within the chamber.

Sensors

Monitoring of relative humidity in controlled environments has been the subject of considerable uncertainty because of the difficulty in obtaining precise measurements. Many continuously operating sensors change calibration over time so that continual recalibration with a precision instrument is required. Most sensors have a response time of one minute or more so that they cannot accurately trace the rapid changes in humidity during a heating and cooling cycle in the chambers or provide an accurate reading at any particular instant.

Psychrometers. The cheapest and one of the most precise methods of determining the moisture level of the chamber atmosphere, the psychrometer consists of two temperature sensitive elements (thermometers, thermocouples, or resistance elements), one of which is covered with a moistened wick. Evaporation of water from the wet element cools it, and the cooling is proportional to the vapor pressure deficit of the air. The wet-bulb temperature, when compared to dry-bulb temperature, can be converted to relative humidity (Table 3.5). Air must move over the wet and dry sensors of the psychrometer at a rate of at least 3 meters per second. Insufficient cooling of the wet bulb will result if the air movement over the sensors is too low. Optimally, the psychrometer should allow temperatures to be read from outside the chamber. Thus, the use of hand-whirled thermometers (sling psychrometers) is not recommended. Battery-powered psychrometers are available for less than $100. They can be started and placed within the chamber and temperatures can be observed through the window of the chamber (Fig. 3.3). Opening the chamber door affects the internal humidity, which will stabilize again after about 10 minutes. The investigator should recognize, however, that precise readings cannot be obtained at any particular intervals. The rapid fluctuations in temperature during heating and cooling cycles do not permit accurate determinations. The large mass of the thermometer bulb in this type of psychrometer causes a slow temperature response and the thermometer cannot reach equilibrium. There also tends to be a difference in time required for temperature equilibrium between the two thermometers, thus the determined relative humidity level should be based on several wet and dry bulb readings taken over one or more complete heating and cooling cycles rather than on any one instantaneous reading.

Psychrometers are subject to additional errors when (1) the cloth on the wet bulb is dirty or oily, (2) the water used is not distilled, or (3) the cloth does not extend far enough beyond the sensing part of the element. A discussion of these errors is provided by Platt and Griffiths (1964).

Hygroscopic elements. Hygroscopic elements provide a reasonably inexpensive and effective method for humidity control. These elements are either human hair or nylon strips that expand and contract with changes in moisture absorption (Fig. 3.4). Thus the element can open or close a relay as it expands and contracts. The elements should be kept clean and brushed with a camelhair brush soaked in distilled water.

The calibration of these elements changes rapidly, particularly if large changes in relative humidity are programmed in

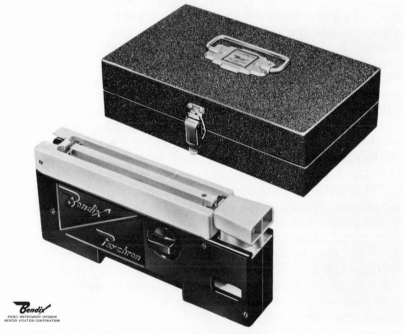

Fig. 3.3. Battery-powered thermometer psychorometer. (Courtesy Environmental Science Division of Bendix.)

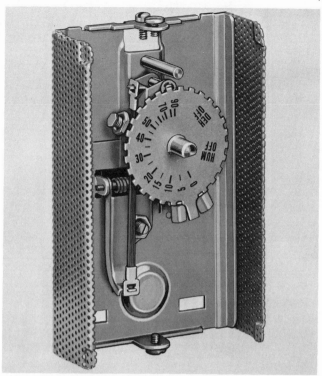

Fig. 3.4. Humidistat with hygroscopic (hair) elements. (Courtesy of Penn Controls.)

the chamber. The calibration should be checked every few days with a precision instrument to insure uniformity.

Animal membranes known as "gold beater's skin" are often employed in the stand-up hygrometers. These are also subject to calibration changes and the dial graduations are rarely accurate.

Electrical conductivity elements. Electrical conductivity elements are commonly used in chambers requiring rapid and precise humidity control (Fig. 3.5). The elements are either flat or tubular and operate by the principle of varying electrical conductivity of a hygroscopic salt, such as lithium chloride, coated between conductive wires wound on the element. Moisture absorption by the salt coating increases the electrical conductivity between the separate strands of the grid. The elements are sensitive to small humidity changes. Some manufacturers provide elements with narrow (20–30%) relative humidity ranges. They must be corrected for the temperature of operation to obtain the desired humidity level. The response of all conductivity elements is not completely linear and the calibration of the elements changes with time and with large humidity fluctuations. The linearity of the elements tends to become poorer with continued use and thus they need to be replaced after 1 to 3 years of use. Free water, as condensation or spray droplets, should not be allowed to deposit on these elements for the calibration will change drastically. In certain elements the foil separates from the base if free water is deposited on it.

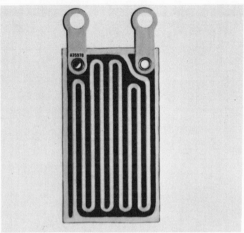

Fig. 3.5. Electrical conductivity element. (Courtesy of Phys-Chemical Research Corp.)

Dew-point apparatus. Instruments for monitoring dew-point temperatures are excellent for continuous humidity measurement. They are particularly useful for calibration of other sensors. The cost of dew-point instruments is in excess of $800. The instruments have a small detector that is cooled as air is passed over it until condensation occurs (Fig. 3.6). The temperature at which condensation occurs is the dew point. Condensation is detected either by an optical system that senses a change in the reflectance of a mirrored surface or by an electrical system that measures the conductivity of an absorbent material. The instruments will stabilize at a new humidity level in less than a minute.

Dew-point sensors can be positioned outside of the controlled chamber and an air sample can be pulled through the instrument. Thus several small chambers can be attached to the same dew-point instrument and monitored consecutively. Care must be

Fig. 3.6. Dew-point hygrometer. (Courtesy of Environmental Equipment Division of EG&G.)

taken that the air sample in the tubing leading to the detector is maintained at a temperature above its dew point to avoid condensation in the lines.

Infrared analyzers. Infrared analyzers are the most precise and accurate of the instruments for monitoring water vapor described in this chapter, and they provide a rapid response (Fig. 3.7). The cost of these instruments ($2000 to $3000), however, has limited their use. The detectors are large and thus primarily suited to remote operation on an air sample drawn from the chamber through the instrument. Infrared detectors require considerable maintenance and the calibration must be continually checked to insure precision.

Procedures for Measurement

The measurement of water vapor in a growth chamber is subject to considerable variation and must be carefully conducted if precise measurements are to be obtained (Wexler 1965). The location of sensors within a chamber is not as critical as with temperature and light sensors, for moisture gradients are commonly not as large. It is desirable to sample the air near the plants, and if wet and dry bulb sensors are utilized, both temperatures must be taken at the same locations.

Fig. 3.7. Infrared analyzer. (Courtesy of Process Instruments Division of Beckman Instruments, Inc.)

Measurements will not be accurate if the door of the chamber cannot be closed while readings are being taken. After the door is closed, *at least ten minutes must elapse for chamber equilibrium before readings are taken.*

The response time of most humidity sensors is quite long, 30 seconds or more, so that control systems cannot respond rapidly to humidification and dehumidification systems and it is difficult to monitor precisely the humidity fluctuations caused by rapid cycling of the refrigeration system.

Most control and recording sensors are subject to changes in calibration, particularly if the chambers are run near saturation during part of the diurnal cycle. Thus it is desirable to check chamber calibration with a precise measuring instrument *daily* during critical environmental experiments.

Humidification

This section describes maintenance requirements for humidifiers provided on commercial chambers and gives information about supplementary humidification equipment.

Humidification in chambers utilizes large quantities of water, many liters per day if significant increases are desired. Humidification should be carried out with deionized water to avoid the deposit of dissolved salts on the chamber surfaces. The use of deionized water also greatly prolongs the life of humidification equipment by avoiding salt buildup.

Spray.   Spray nozzles that direct a fine mist of water into the air stream are provided in some chambers. The fine droplets evaporate in the air stream increasing the water vapor in the air. Evaporation is most effective when the spray is directed perpendicular to or directly against the air stream so that the transport of large water droplets in the air stream is minimized. Water droplets do not evaporate and may be deposited on leaf surfaces and chamber surfaces, especially when the chamber's atmosphere is maintained close to saturation. The nozzles wear out rapidly and must be adjusted and replaced regularly to maintain a fine mist. Water pressure of 20 to 60 psi must be provided to insure proper misting.

Atomizers.   Several chamber manufacturers provide portable atomizers for humidification (Fig. 3.8). These atomizers have the advantage of being readily moved around within the chamber, though it is difficult to position them so that uniform distribution of moisture is obtained.

Steam.   Steam for humidification is commonly recommended when a supply is available, and it is an effective means of humidification. The steam must be free of additives and properly

Fig. 3.8. Portable atomizer for humidification. (Courtesy of Herrmidifier Company, Inc.)

distributed in the chamber. It has the advantage of providing moisture free of dissolved salts. Steam has to be admitted so as to avoid excessive heating in portions of the chamber where it is directed. If steam is obtained from institutional heating lines there may be problems from chemicals added for fungal, bacterial, and pH control in the condensate return lines. Chemicals commonly used include cyclohexylamine, morpholine, and actadecylamine, which have phytotoxic activity in low concentration, particularly cyclohexylamine. (See Chapter 6, Air Contaminants.)

Water baths. Some commercial chambers provide water baths for humidification. Air is blown over the bath and then directed into the chamber. A thermostatically controlled heater regulates the temperature of the water and thus increases or decreases the amount of humidification in the chambers. The amount of humidification provided by a water bath is a function of the water surface exposed, the water temperature, and the air turbulence over the bath. Humidity levels close to 100 percent usually cannot be obtained during the light period with a water bath system because of the reduced evaporation from the water baths as the air nears saturation. Increasing the temperature of the water bath has limitations for increasing humidity, for as the temperature of the water bath is increased, the heat added to the air is increased. This necessitates additional cooling in the chambers, which in turn tends to decrease the moisture levels in the chamber.

The water bath should be provided with distilled water to minimize salt accumulation in the tank. If tap water has to be used, a regular dilution or bleeding of excess water from the tank must be instituted to avoid salt accumulation. The level of water in the bath is commonly maintained by a float system, which must be adjusted to compensate for differences in water level when the humidifier fan is switched on and off. It is also desirable to have a temperature cutoff on the water bath heater if the supply of water is disrupted.

Saturated salt and sulphuric acid solutions. Theoretically, the use of saturated salt and sulphuric acid solutions should maintain the moisture in a chamber at an established level for the particular salt or acid solution. However, the large requirements for water vapor in chambers cannot be effectively met with these solutions, and they have not been found to be useful in growth chambers.

Dehumidification. The problem of dehumidification in controlled environment chambers is met more easily than humidification unless very arid conditions or low temperatures are desired. The normal operation of a chamber constantly provides a degree of dehumidification as moisture is condensed on the cold coils. The amount of cooling and hence the rate of dehumidification increase with increasing light or other heat loads in the chamber.

As indicated previously, humidity levels in conventional chambers with an on-off cycling of the refrigeration system during the light period will be about 50 percent RH for temperatures of 15° to 30°C. This will increase to near saturation during the dark period unless dark-period heaters are used or the temperature of the building is above the temperature being maintained within the chamber. The presence of plants in a chamber will not significantly influence humidity levels during the light period unless extremely arid conditions are required. The presence of plants will, however, increase humidity levels during the dark period unless the heat load during the dark is quite large.

Table 3.1. **Density of water vapor at saturation over water (atmospheric pressure = 29.92 in, 760 mm, 1013 mb)**

°C	$g \cdot m^{-3}$	°C	$g \cdot m^{-3}$	°C	$g \cdot m^{-3}$
0	4.84	17	14.48	34	37.61
1	5.19	18	15.37		
2	5.55	19	16.31	35	39.63
3	5.94			36	41.75
4	6.36	20	17.30	37	43.96
		21	18.34	38	46.26
5	6.79	22	19.43	39	48.67
6	7.26	23	20.58		
7	7.75	24	21.78	40	51.19
8	8.27			41	53.82
9	8.81	25	23.05	42	56.56
		26	24.38	43	59.41
10	9.33	27	25.78	44	62.39
11	10.01	28	27.24		
12	10.66	29	28.78	45	65.50
13	11.35			46	68.73
14	12.07	30	30.38	47	72.10
		31	32.07	48	75.61
15	12.83	32	33.83	49	79.26
16	13.63	33	35.68	50	83.06

Source: Smithsonian Meteorological Tables, 6th rev. ed. 1951.

Low levels of humidity (10–50% RH) are provided most commonly by separate dehumidification systems affixed to the chambers. These contain rechargeable or replaceable dessicants and the amount of dehumidification obtained is dependent upon the amount of chamber air passed through the units in a given time and the effectiveness of the desiccants. Dehumidification systems utilizing desiccants have limited usefulness when moisture additions are so high that the desiccant capacity is exhausted rapidly.

Table 3.2. **Saturation vapor pressure deficits**

Temperature	Relative humidity									
°C	0	10	20	30	40	50	60	70	80	90
	Millibars of pressure*									
0	6.11	5.50	4.89	4.28	3.67	3.05	2.44	1.83	1.22	.61
5	8.72	7.85	6.98	6.10	5.23	4.36	3.49	2.62	1.74	.87
10	12.27	11.04	9.82	8.59	7.36	6.14	4.91	3.68	2.45	1.23
15	17.04	15.34	13.64	11.93	10.23	8.52	6.82	5.11	3.41	1.70
20	23.37	21.04	18.70	16.36	14.02	11.69	9.35	7.01	4.68	2.34
25	31.67	28.50	25.34	22.17	19.00	15.84	12.67	9.50	6.33	3.17
30	42.43	38.19	33.94	29.70	25.46	21.22	16.97	12.73	8.49	4.24
35	56.24	50.61	44.99	39.37	33.74	28.12	22.49	16.87	11.25	5.62
40	73.78	66.40	59.02	51.64	44.27	36.89	29.51	22.13	14.76	7.39
45	95.86	86.27	76.68	67.10	57.51	47.93	38.34	28.76	19.17	9.59
50	123.40	111.06	98.72	86.38	74.04	61.70	49.36	37.02	24.68	12.34

Source: Smithsonian Meteorological Tables, 6th rev. ed. 1951.

*Conversion to millimeters: multiply millibars by 0.752. Conversion to inches: multiply millibars by .039.

***Table* 3.3. Conversion of dew-point temperatures to relative humidity (atmospheric pressure = 760 mm, 29.92 in, 1013 mb)**

Depression of dew point °C (t−d)	Dew point (d) °C					Depression of dew point °C (t−d)	Dew point (d) °C					Depression of dew point °C (t−d)	Dew point (d) °C			
	0	10	20	30	40		0	10	20	30	40		0	10	20	30
0.0	100	100	100	100	100	5.4	68	70	72	74	75	12.0	44	47	49	52
0.2	99	99	99	99	99	5.6	67	69	71	73	75	12.5	42	45	48	50
0.4	97	97	98	98	98	5.8	66	69	70	72	74	13.0	41	44	46	49
0.6	96	96	96	97	97	6.0	66	68	70	71	73	13.5	40	43	45	48
0.8	94	95	95	96	96	6.2	65	67	69	71	72	14.0	38	41	44	47
1.0	93	94	94	94	95	6.4	64	66	68	70	72	14.5	37	40	43	45
1.2	92	92	93	93	94	6.6	63	65	67	69	71	15.0	36	39	42	44
1.4	90	91	92	92	93	6.8	62	64	66	68	70	15.5	35	38	40	43
1.6	89	90	91	91	92	7.0	61	63	66	68	70	16.0	34	37	39	42
1.8	88	89	90	90	91	7.2	60	63	65	67	69	16.5	33	36	38	41
2.0	87	88	88	89	90	7.4	60	62	64	66	68	17.0	32	35	37	40
2.2	85	86	87	88	89	7.6	59	61	63	65	68	17.5	31	34	36	39
2.4	84	85	86	87	88	7.8	58	60	63	65	67	18.0	30	33	35	38
2.6	83	84	85	86	87	8.0	57	60	62	64	66	18.5	29	32	34	37
2.8	82	83	84	85	86	8.2	56	59	61	63	66	19.0	28	31	33	36
3.0	81	82	83	84	85	8.4	56	58	60	63	65	19.5	27	30	33	35
3.2	80	81	82	83	85	8.6	55	57	60	62	64	20.0	26	29	32	34
3.4	79	80	81	82	84	8.8	54	57	59	61	64	21.0	25	27	30	—
3.6	77	79	80	81	83	9.0	53	56	58	61	63	22.0	23	26	28	—
3.8	76	78	79	81	82	9.2	53	55	58	60	62	23.0	22	24	27	—
4.0	75	77	78	80	81	9.4	52	55	57	59	62	24.0	21	23	26	—
4.2	74	76	77	79	80	9.6	51	54	56	59	61	25.0	19	22	24	—
4.4	73	75	77	78	79	9.8	51	53	56	58	60	26.0	18	21	23	—
4.6	72	74	76	77	78	10.0	50	53	55	57		27.0	17	20	22	—
4.8	71	73	75	76	78	10.5	48	51	54	56		28.0	16	19	21	—
5.0	70	72	74	75	77	11.0	47	49	52	55		29.0	15	18	20	—
5.2	69	71	73	75	76	11.5	45	48	51	53		30.0	14	17	19	—

Table 3.4. **Conversion of relative humidity to dew-point temperature (atmospheric pressure = 760 mm, 29.92 in, 1013 mb)**

Dry-bulb temperature °C	Relative humidity								
	10	20	30	40	50	60	70	80	90
0	-25.4	-18.3	-14.0	-10.8	-8.3	-6.1	-4.3	-2.6	-1.3
	(25.4)*	(18.3)	(14.0)	(10.8)	(8.3)	(6.1)	(4.3)	(2.6)	(1.3)
5	-21.2	-14.6	-9.9	-6.7	-4.1	-2.0	-0.1	+1.8	+3.5
	(26.2)	(19.6)	(14.8)	(11.7)	(9.1)	(7.0)	(4.9)	(3.2)	(1.5)
10	-18.5	-10.8	-5.9	-2.6	-0.0	+2.5	+4.8	+6.8	+8.5
	(28.5)	(20.8)	(15.9)	(12.6)	(10.0)	(7.5)	(5.2)	(3.2)	(1.5)
15	-14.6	-6.9	-2.2	-1.5	+4.7	+7.4	+9.7	+11.6	+13.4
	(29.6)	(21.9)	(17.2)	(13.5)	(10.3)	(7.6)	(5.3)	(3.4)	(1.6)
20	-11.0	-3.3	+2.0	+6.0	+9.3	+12.0	+14.4	+16.5	+18.3
	(31.0)	(23.3)	(18.0)	(14.0)	(10.7)	(8.0)	(5.6)	(3.5)	(1.7)
25	-7.7	+0.6	+6.2	+10.5	+13.9	+16.8	+19.1	+21.3	+23.2
	(32.7)	(24.4)	(18.8)	(14.5)	(11.1)	(8.2)	(5.9)	(3.7)	(1.8)
30	-4.3	+4.7	+10.7	+14.9	+18.4	+21.4	+23.9	+26.2	+28.2
	(34.3)	(25.3)	(19.3)	(15.1)	(11.6)	(8.6)	(6.1)	(3.8)	(1.8)
35	-1.0	+8.7	+14.9	+19.4	+23.0	+26.1	+28.7	+31.0	+33.1
	(36.0)	(26.3)	(20.1)	(15.6)	(12.0)	(8.9)	(6.3)	(4.0)	(1.9)
40	+2.5	+12.9	+19.1	+23.8	+27.6	+30.7	+33.5	+35.8	+38.0
	(37.5)	(27.1)	(20.9)	(16.2)	(12.4)	(9.3)	(6.5)	(4.2)	(2.0)
45	+6.4	+16.9	+23.4	+28.2	+32.1	+35.4	+38.2	+40.7	+43.0
	(38.6)	(28.1)	(21.6)	(16.8)	(12.9)	(9.6)	(6.8)	(4.3)	(2.0)
50	+10.1	+20.8	+27.6	+32.7	+36.7	+40.1	+43.0	+45.6	+47.9
	(39.9)	(29.2)	(22.4)	(17.3)	(13.3)	(9.9)	(7.0)	(4.4)	(2.1)

Source: Adapted from Smithsonian Meteorological Tables, 6th rev. ed. 1951.
*Dew-point temperature depressions in parenthesis.

Table 3.5. **Conversion of wet- and dry-bulb temperatures to relative humidity (atmospheric pressure = 760 mm, 29.92 in, 1013 mb)**

Depression of wet-bulb thermometer °C (t-t')	Dry-bulb temperature °C (t)														
	0	1	2	3	4	5	6	7	8	9	10	11	12	13	14
.5	91	91	92	92	92	93	93	93	94	94	94	94	94	95	95
1.0	82	83	84	84	85	86	86	87	87	88	88	88	89	89	90
1.5	73	75	76	77	78	79	79	80	81	82	82	83	83	84	84
2.0	65	66	68	69	70	72	73	74	75	76	76	77	78	79	79
2.5	56	58	60	62	63	65	66	67	69	70	71	72	73	74	74
3.0	48	49	52	54	56	58	60	61	63	64	65	66	68	69	70
3.5	39	42	45	47	49	51	53	55	57	58	60	61	62	64	65
4.0	31	34	37	40	42	45	47	49	51	53	54	56	57	59	60
4.5	21	25	30	33	36	38	41	43	45	47	49	51	53	54	56
5.0	13	17	22	26	29	32	35	37	40	42	44	46	48	49	51
5.5	5	10	14	19	22	26	29	31	34	36	39	41	43	45	46
6.0	—	—	7	12	16	20	23	26	29	31	34	36	39	41	42
6.5			—	5	9	13	17	20	23	26	29	31	33	38	38
7.0				—	—	7	11	14	18	21	24	27	29	32	33
7.5					—	5	10	14	17	29	22	24	27	39
8.0						—	—	8	12	14	17	20	23	25
8.5							—	7	10	13	16	18	21
9.0								—	6	9	12	14	17
9.5									—	5	8	11	14
10.0										—	—	7	10

Table 3.5. (cont.) **Conversion of wet- and dry-bulb temperatures to relative humidity (atmospheric pressure = 760 mm, 29.92 in, 1013 mb)**

Depression of wet-bulb thermometer °C (t-t')	Dry-bulb temperature °C (t)															
	15	16	17	18	19	20	21	22	23	24	25	26	27	28	29	30
.5	95	95	95	95	95	96	96	96	96	96	96	96	96	96	96	96
1.0	90	90	90	91	91	91	91	92	92	92	92	92	92	93	93	93
1.5	85	85	86	86	86	87	87	88	88	88	88	88	89	89	89	89
2.0	80	81	81	82	82	83	83	84	84	84	84	85	85	85	86	86
2.5	75	76	77	77	78	78	79	80	80	80	81	81	81	82	82	83
3.0	71	71	72	73	74	74	75	76	76	77	77	78	78	78	79	79
3.5	66	67	68	69	70	70	71	72	72	73	74	74	75	75	76	76
4.0	61	63	64	65	66	66	67	68	69	69	70	71	71	72	72	73
4.5	57	58	59	61	62	62	64	64	65	66	67	67	68	68	69	70
5.0	52	54	55	56	58	59	60	61	62	63	63	64	65	65	66	67
5.5	48	50	51	53	54	55	56	57	58	59	60	61	62	62	63	64
6.0	44	46	47	49	50	51	53	54	55	56	57	58	58	59	60	61
6.5	40	42	43	45	46	48	49	50	51	53	54	55	55	56	57	58
7.0	36	37	39	41	43	44	46	47	48	49	50	52	52	53	54	55
7.5	32	34	36	37	39	41	42	44	45	46	47	48	50	50	51	52
8.0	27	30	32	34	36	37	39	40	42	43	44	46	47	48	49	50
8.5	23	26	28	30	32	34	36	37	39	40	41	43	44	45	46	47
9.0	20	22	24	27	29	30	32	34	36	37	38	40	41	42	43	44
9.5	16	18	21	23	25	27	29	31	33	34	36	37	38	40	41	42
10.0	12	15	17	20	22	24	26	28	30	31	33	34	36	37	38	39
11.0	6	8	10	13	15	18	20	22	24	26	27	29	30	32	33	35
12.0	—	—	4	7	10	12	14	16	18	20	22	24	25	27	28	30
13.0				—	3	6	8	11	13	15	17	19	21	22	24	25
14.0					—	3	5	8	10	12	14	16	18	19	21	
15.0							—	3	5	7	9	11	13	15	17	
16.0									—	3	5	7	9	11	13	
17.0											—	3	5	7	9	
18.0														1	3	5
19.0																1

Table 3.5. (cont.) **Conversion of wet- and dry-bulb temperatures to relative humidity (atmospheric pressure = 760 mm, 29.92 in, 1013 mb)**

Depression of wet-bulb thermometer °C (t-t')	Dry-bulb temperature °C (t)													
	31	32	33	34	35	36	37	38	39	40	41	42	43	44
.5	96	96	97	97	97	97	97	97	97	97	97	97	97	97
1.0	93	93	93	93	94	94	94	94	94	94	94	94	94	94
1.5	90	90	90	90	90	90	91	91	91	91	91	91	91	91
2.0	86	86	87	87	87	87	87	87	88	88	88	88	88	89
2.5	83	83	83	84	84	84	84	84	85	85	85	86	86	86
3.0	80	80	80	81	81	81	82	82	82	82	83	83	83	83
3.5	77	77	77	78	78	78	79	79	79	80	80	80	80	81
4.0	73	74	74	75	75	75	76	76	77	77	77	78	78	78
4.5	70	71	71	72	72	73	73	74	74	74	75	75	75	76
5.0	67	68	69	69	69	70	70	71	71	72	72	72	73	73
5.5	64	65	66	66	67	67	68	68	69	69	70	70	70	71
6.0	62	62	63	64	64	64	65	66	66	67	67	67	68	68
6.5	59	60	60	61	61	62	63	63	64	64	65	65	66	66
7.0	56	57	58	58	59	59	60	61	61	62	62	63	63	64
7.5	53	54	55	56	56	57	58	58	59	59	60	61	61	62
8.0	50	51	52	53	54	55	55	56	57	57	58	58	59	59
8.5	48	49	50	51	51	52	53	54	54	54	56	56	57	57
9.0	45	46	47	48	49	50	51	51	52	53	53	54	55	55
9.5	43	44	45	46	47	48	48	49	50	51	51	52	52	53
10.0	40	41	42	43	44	45	46	47	48	48	49	50	50	51
11.0	36	37	38	39	40	41	42	43	43	44	45	46	46	47
12.0	31	32	33	34	36	37	38	39	39	40	41	42	43	43
13.0	27	28	29	30	31	32	34	35	36	36	37	38	39	40
14.0	22	24	25	26	27	29	30	31	32	33	34	35	35	36
15.0	18	20	21	22	24	25	26	27	28	29	30	31	32	33
16.0	14	16	17	19	20	21	22	24	25	26	27	28	29	30
17.0	10	12	14	15	16	18	19	20	21	23	24	25	26	27
18.0	7	8	10	12	13	14	16	17	18	20	21	22	23	24
19.0	3	5	7	8	10	11	13	14	15	17	18	19	20	21
20.0		—	—	5	7	8	10	11	12	14	15	16	17	18
21.0				—	4	5	7	8	10	11	13	13	14	16
22.0					—	4	5	7	8	9	11	12	13
23.0							—	—	6	7	8	9	10
24.0									—	—	6	7	8
25.0											—	5	6
26.0												—	—

Table 3.6. **Correction of relative humidity for atmospheric pressure difference of 10 mm**

Air temperature °C	Relative humidity (%)								
	10	20	30	40	50	60	70	80	90
0	.76	.66	.59	.50	.40	.32	.25	.15	.07
5	.68	.61	.54	.45	.37	.29	.22	.14	.07
10	.62	.54	.47	.39	.33	.25	.19	.12	.06
15	.54	.46	.40	.32	.28	.22	.16	.10	.05
20	.47	.41	.35	.29	.24	.19	.13	.09	.04
25	.41	.35	.30	.24	.20	.15	.11	.06	.03
30	.35	.30	.26	.21	.17	.13	.09	.06	.03
35	.29	.24	.20	.17	.13	.10	.07	.05	.02
40	.26	.22	.18	.14	.12	.09	.06	.04	.02

Source: Adapted from Konig, Preuss. Meteorl. Inst. Tables. 1914.

Note: For each 10 mm of increased pressure the correction is subtracted. For each 10 mm of decreased pressure the correction is added.

References

Ford, M. A., and G. N. Thorne. 1974. Effects of atmospheric humidity on plant growth. *Ann. Bot.* 38:441–452.

Gaastra, P. 1959. Photosynthesis of crop plants as influenced by light, carbon dioxide, temperature, and stomatal diffusion resistance. *Meded. Landbouhoogesch. Wageningen* 59 (13):1–68.

Hoffman, G. J. 1972. Humidity effects on yield and water relations of nine crops. In *Proc. UNESCO:NSF Phytotron Symposium,* May 1972. Raleigh, N.Y.

Hoffman, G. J., and S. L. Rawlins. 1971. Growth and water potential of root crops as influenced by salinity and relative humidity. *Agron. J.* 63:877–880.

Hoffman, G. J.; S. L. Rawlins; M. J. Garber; and E. M. Cullen. 1971. Water relations and growth of cotton as influenced by salinity and relative humidity. *Agron. J.* 63:822–826.

Kozlowski, T. T., ed. 1976. *Water Deficits and Plant Growth IV.* Academic Press, New York. 383 pp.

Kristoffersen, T. 1963. Interactions of photoperiod and temperature in growth and development of young tomato plants. *Physiol. Plant.* (Suppl. 1). 98 pp.

Krizek, D. T.; W. A. Bailey; and H. H. Klueter. 1971. Effects of relative humidity and type of container on the growth of F_1 hybrid annuals in controlled environments. *Amer. J. Bot.* 58(6):544–551.

Matsui, T., and H. Eguchi. 1972. Effects of environmental factors on leaf temperature in a temperature controlled room. *Environ. Control in Biol.* 10:15–18.

Meidner, H., and T. A. Mansfield. 1968. *Physiology of Stomata.* McGraw-Hill, New York. 178 pp.

Palzkill, D. A.; T. W. Tibbitts; and P. H. Williams. 1976. Enhancement of calcium transport to inner leaves of cabbage for prevention of tipburn. *J. Amer. Soc. Hort. Sci.* 101:645–648.

Platt, R. B., and J. F. Griffiths. 1964. *Environmental Measurement and Interpretation.* Reinhold, New York. 235 pp.

Read, M. 1972. Growth and tipburn of lettuce: Carbon dioxide enrichment at different light intensity and humidity levels and rate of incorporation of carbon-14 assimilates into the latex. Ph.D. thesis, University of Wisconsin, Madison.

Slayter, R. O. 1973. Plant response to climatic factors. In *Proceedings of 1970 Uppsala Symposium*. UNESCO, Paris.

Tibbitts, T. W., and G. Bottenberg. 1976. Growth of lettuce under controlled humidity levels. *J. Amer. Soc. Hort. Sci.* 101:70–73.

Tromp, J., and J. Oele. 1972. Shoot growth and mineral composition of leaves of fruits of apple as affected by relative air humidity. *Physiol. Plant.* 27:253–258.

Troughton, J. H., and R. O. Slayter. 1969. Plant water status, leaf temperature, and the calculated mesophyll resistance to carbon dioxide of cotton leaves. *Aust. J. Biol. Sci.* 22:815–828.

Wexler, A., ed. 1965. *Humidity and Moisture: Measurement and Control in Science and Industry*. Vol. 1. *Principles and Methods of Measuring Humidity in Gases*, Ed. R. E. Ruskin. Vol. 3. *Fundamentals and Standards*, Ed. A. Wexler and W. A. Wildhack, Reinhold, New York.

Carbon Dioxide

Carbon dioxide (CO_2), a normal constituent of the air of vital importance to plant growth and development, is assimilated by plants in the presence of radiation with wavelengths between 400 and 700 nm to produce sugars. This significant plant process stores the energy, in the form of carbohydrates, upon which most life on this planet is dependent. Carbon dioxide is released back to the atmosphere as the carbohydrates are metabolized by plants and animals or in the combustion of fossilized forms of carbohydrates as coal and oil. Carbon dioxide in the atmosphere absorbs long wave radiation that maintains the particular heat balance necessary for life on this planet.

The natural level of CO_2 in air is not static. Through the combustion of fossil fuels, CO_2 levels have increased appreciably since the turn of the century from about 290 ppm (0.029%) to about 320 ppm (0.032%) (Amer. Chem. Soc. 1969). The increase is continuing at a rate of about 0.7 ppm per year.

In air polluted by industry, traffic, heating, or volcanic activity the CO_2 level in the atmosphere may rise to 600 ppm or higher (Pallas 1970). The average level in large urban areas will be at least 10 percent greater than in rural areas. Normal human respiration is a significant source of CO_2 enhancement in buildings and confined areas. The average CO_2 concentration in the home is about 600 ppm. An adult person at rest exhales about 300 liters of air per hour; the exhaled air contains 4 to 5 percent CO_2. Thus, the purported benefit of "talking to your plants" may well be that of CO_2 enrichment.

On the other hand, photosynthesis removes CO_2 from the air.

The resulting decreases were measured both in field conditions in solid canopies of plants (Lemon 1970) and in greenhouses when vents were closed to maintain desirable temperature conditions (Wittwer and Robb 1964, Holley 1965, Bierhuizen 1973, Chapman et al. 1954, Pearman and Garratt 1973).

Plants respond to increased CO_2 levels with increased photosynthetic rates; however, humans do not benefit from high CO_2 levels. At CO_2 levels above 2 percent, injury can result to both plants and animals. The risk is more serious for animals, and particularly humans, than for plants; a short exposure to 8 percent CO_2 constitutes an acute danger to human life.

Controlled Environments

In recent years, the carbon dioxide level in controlled environments has been recognized as one of the environmental parameters that must be maintained for effective plant research. It has been generally assumed that small fans supplying fresh air to chambers were adequate to maintain satisfactory CO_2 levels for plant growth. Thus, provision for monitoring CO_2 and control systems to supply additional CO_2 have not been included in design criteria for standard plant growth chambers. With careful measurements of CO_2 levels in controlled environments, however, it has become clear that the amount of CO_2 varies significantly. If CO_2 levels are naturally high in certain instances growth may be rapid, and if unnaturally low growth may be seriously limited.

All chambers are subject to CO_2 fluctuations, principally caused by human and plant activity. Thus, whenever a human walks into a chamber and exhales air that is 4 to 5 percent CO_2, the amount of CO_2 in the chamber may increase as much as tenfold, far above the normal atmospheric level of .03 percent. After the human leaves the chamber, the level of CO_2 is gradually lowered by dilution, dependent upon the rate of fresh air exchange, or by photosynthetic incorporation, dependent upon the amount of plant tissue in the chamber.

Plant activity has different effects during the light and dark periods. During the light period, photosynthesis by the plant decreases CO_2 levels, as demonstrated by the curve in Figure 4.1. This decrease can be very rapid, causing limiting growth conditions within 15 minutes and essentially stopping CO_2 incorporation in 30 minutes. At night when photosynthesis has ceased, respiration continues and CO_2 released into the chamber raises the CO_2 level above ambient. Plant respiration will more than double CO_2 concentrations during a dark period if the chamber is sealed.

Problems of varying levels of CO_2 are most acute in built-in rooms that have very little leakage and may be totally dependent upon "fresh air" intake fans. On the other hand, commercial chambers have fewer problems because the leakage through joints and connections provides air exchange in addition to the "fresh air" intake. In one new reach-in chamber the leakage was equal to 30 percent of the internal volume in five minutes even though the fresh air intake fans were sealed off. Therefore, the carbon dioxide level of the room in which chambers are placed is of as much concern as the carbon dioxide level in the controlled environments themselves. Location of chambers in areas that have a lot of human activity and a minimum of fresh air exchange will lead to very large fluctuations in CO_2. The fluctuations will depend primarily on the number of hours of human activity in the room. Growth conditions may be affected on certain days if human activity varies or if the light period of separate chambers is staggered.

Regardless of the amount of fresh air added to a chamber, there will be some CO_2 depression during the light period and an increase during the dark period that automatically introduces a CO_2 cycle in combination with the light-dark cycle. The magnitude of this fluctuation is shown in Figure 4.1 utilizing data obtained for a reach-in chamber filled with butterhead lettuce at 20°C, 80%RH and light intensity of approximately 32.5 nE cm^{-2}s^{-1} (2000 ft-c). Table 4.1 has been calculated for different amounts of fresh air exchanged in the chamber.

Chemical and Physical Properties

Carbon dioxide (or carbonic anhydride) is a nonflammable, colorless, odorless gas at room temperature, and a volatile, colorless liquid or a white snowlike solid subliming at −78.5°C. It has a molecular weight of 44.01 and is approximately one and one-half times as heavy as air. Carbon dioxide comprises approximately 0.03 percent of dry air by volume and approximately 0.05 percent by weight. One volume of CO_2 will dissolve in approximately one volume of water at atmospheric pressure and 15°C. Its solubility in pure water at 0°C is twice that at 20°C and nearly three times

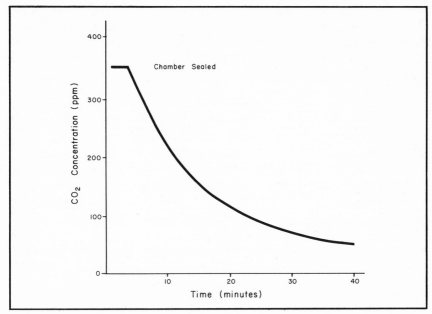

Fig. 4.1. Carbon dioxide concentration in atmosphere of growth chamber with shelf covered with leaf lettuce under approximately 32.5 nE cm^{-2}s^{-1}, 20°C, and 80% RH, plant-bed surface of 39 dm² and chamber volume of 675 L.

Table 4.1. **Carbon dioxide levels in a reach-in chamber affected by fresh air exchange**

Additions of fresh air of 320 ppm CO_2 (chamber volumes per minute)	Equilibrium CO_2 level (ppm)	
	*Light**	*Dark***
1.0	313	324
.9	312	324
.8	311	325
.7	310	325
.6	308	326
.5	306	327
.4	302	329
.3	296	332
.2	284	338
.1	248	356

*Net photosynthetic rate of 13.7 mg CO_2 hr^{-1} dm^{-2} plant-bed surface area butterhead lettuce at 20°C, 80% RH, and 32.5 nE cm^{-2} s^{-1}, plant-bed surface of 39 dm².

**Net respiration rate of 6.8 mg CO_2 hr^{-1} dm^{-2} plant-bed surface area butterhead lettuce at 20°C and 80% RH.

that at 30°C (Table 4.2) (Šesták et al. 1971). Under normal conditions encountered in the laboratory, carbon dioxide is a very stable compound. At temperatures in excess of 1700°C it dissociates slightly into oxygen and carbon monoxide. At high concentrations, CO_2 has an acidic taste.

Table 4.2. **Some properties of carbon dioxide (Atmospheric pressure 760 mm Hg)**

	Temperature °C					
	0	10	15	20	25	30
Density						
kg·m⁻³	1.977	1.907	1.874	1.842	1.811	1.781
m³kg⁻¹	0.506	0.524	0.534	0.543	0.552	0.561
Solubility in pure water						
dm³m⁻³	1713	1194	1019	878	759	665
mol·m⁻³	76.4	51.4	43.1	36.5	31.1	26.7
mg·kg⁻¹	3346	2318	1970	1688	1449	1257

Source: Adapted from Šesták, Čatský, and Jarvis 1971.

When carbon dioxide additions are required in controlled environments, CO_2 is most commonly obtained as a compressed gas in cylinders, although it can be obtained in a frozen state or generated from a salt through addition of acid. Carbon dioxide is provided to greenhouses through combustion of methane or propane but this method is discouraged strongly in controlled environments because of the difficulties of obtaining complete combustion and the danger of toxic gas buildup.

Most laboratory supply companies carry four grades of carbon dioxide: research grade (99.995% minimum purity); instrument grade (99.99% minimum purity); bone dry grade (99.8% minimum purity); and commercial grade (99.5% minimum purity). Typical specifications for the bone dry grade would be nitrogen and oxygen, .05%; dew point, −30°F; and oil content less than 5 ppm. Typical specifications for commercial grade would be nitrogen, .34% maximum; oxygen, .09% maximum; and water, .07% maximum.

For use in CO_2 enrichment studies in plant growth chambers, commercial grade is the most reasonably priced and is quite adequate. For calibration of infrared analyzers, however, instrument or bone dry grade is recommended.

Compressed Gas Cylinders

Dry CO_2 is a relatively inert gas and any common or commercially available metal is used in the construction of CO_2 cylinders. Piping systems should be designed to have a working pressure as specified by a competent engineer using a safety factor conforming to the American Society of Mechanical Engineers code for pressure piping. CO_2 is shipped in Department of Transportation approved cylinders of the high pressure type having a maximum rated service of 1800 psi. CO_2 cylinders are equipped with valves having a Compressed Gas Association valve outlet connection No. 320 with an outlet thread size designated as 0.825 inch–14 threads per inch right-hand external with a flat face seating against a washer (Fig. 4.2A). CO_2 lecture bottles have a special 5/16 inch–32 threads per inch female outlet and 9/16 inch–18 threads per inch male dual-valve outlet (Fig. 4.2B).

The only type of safety device permitted in CO_2 cylinders, for industrial purposes, is the frangible disk. The frangible disk is contained in a special fitting and is an integral part of the cylinder

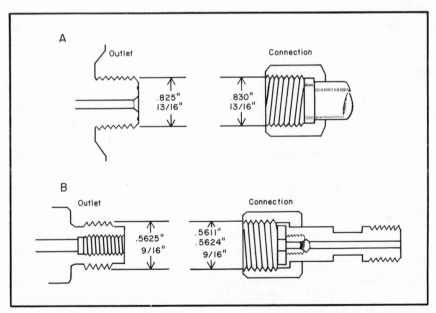

Fig. 4.2. Valve outlets and connections for carbon dioxide compressed gas cylinders. (A) Compressed Gas Association valve outlet connection No. 320. (B) CO_2 lecture bottle outlet. (Braker and Mossman 1971.)

valve. The disk will burst at a pressure not exceeding the minimum required hydrostatic test pressure of the cylinder, which is substantially below the bursting pressure of the cylinder. Bursting of the frangible disk results in discharge of the entire contents of the cylinder.

There are three types of automatic pressure regulators that may be used with compressed gas cylinders—single-stage regulator, two-stage regulator, and low-pressure regulator.

The single-stage regulator reduces cylinder pressure in one step to a range of delivery pressures depending upon the design of the regulator and its spring load. As cylinder pressure falls, a single-stage regulator will show a decrease in delivery pressure. Since CO_2 is a liquefied gas at room temperature, the cylinder pressure will remain reasonably constant as long as there is any liquid CO_2 remaining in the cylinder. Thus, a steady delivery pressure will be produced until approximately 80 percent of the CO_2 in the cylinder has been discharged. Above the critical temperature of 31°C (87.8°F) carbon dioxide converts completely to a gas so the discharge of gas will show a steady drop in pressure. The range in delivery pressures available is from 4–80 to 100–1500 psi.

The two-stage regulator performs the same function as a single-stage regulator but maintains greater accuracy in the control of delivery pressure, which varies very little as cylinder pressure drops. The range in delivery pressures available is from 2–15 to 20–250 psi.

For certain applications requiring a very low delivery pressure (below 10 psi or 69kPa) and a constant delivery pressure, a low-pressure regulator is used in combination with a one or two-stage regulator. In most controlled environment studies, however, this regulator will not be used.

When a needle valve or other regulator valve is used with the regulator, it should be attached directly to the regulator or connected with a high pressure hose.

In order to maintain accurate flow rates, a good flowmeter should be used. When making adjustments of flow on a flowmeter, always adjust from an excess of flow to the desired level rather than from a reduced flow to the desired level to avoid the lag in obtaining an equilibrium flow due to pressure build-up between the pressure regulator and flowmeter.

Manual controls with manual needle valves that connect directly to the cylinder valve outlet are available but are not recommended for use in growth chamber studies. These controls require close supervision and should not be used as pressure controls, since dangerous pressures can develop if a line or system becomes plugged.

Frozen carbon dioxide (dry ice). Use of dry ice for addition of CO_2 to chambers is not recommended because of the difficulties in obtaining a uniform release of CO_2 and the high cost of handling this source.

Salt with acid. Production of CO_2 from salts as sodium carbonate or potassium carbonate with the addition of acid is not recommended for regular use in controlled environment chambers because of the complexities of maintaining the salt-acid system continuously. This system is used for addition of radioactive carbon dioxide to small chambers and might be useful where rapid bursts of carbon dioxide are needed.

Human Toxicity

The problem of CO_2 toxicity is seldom encountered in most growth chamber studies since the level of CO_2 used to enhance plant growth is generally below 2000 ppm (0.2%), considerably below the level (1.5%) at which certain physiological effects can be observed upon man (Roth 1964). The time dependence of toxic levels of CO_2 is shown in Figure 4.3. The dangerous level for humans that will cause dizziness and unconsciousness is above 5 percent CO_2. For prolonged periods of exposure a level below 2 percent CO_2 should be insured (Glatte and Welch 1967).

Influence on Plant Growth

The level of carbon dioxide is of much greater significance and concern in controlled environments than in field environments. The rate of air exchange in the controlled environments is low and large fluctuations in the amount of CO_2 can occur. Increases encourage plant growth and significant depletions may greatly slow or stop plant growth. Several studies have documented the growth resulting from carbon dioxide increases above the ambient level (approximately 320 ppm) (Ford and Thorne 1967, Kretchman and Howlett 1970, Wittwer 1970, Holley 1970, Gaastra 1959, Krizek et al. 1971, Madsen 1973). Growth increases from high CO_2 levels are measured as gains in fresh and dry weight and are visible as more rapid maturation of plants, in-

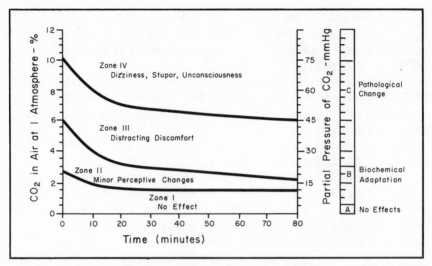

Fig. 4.3. Time dependence of carbon dioxide toxicity to humans. (Glatte and Welch 1967.)

creased branching, and larger leaves. These growth increases are nearly linear with CO_2 levels to 1000 ppm, and continue to rise with CO_2 concentrations up to about 2000 ppm, as shown by Gaastra (1959).

On the other hand, growth slows as the amount of CO_2 is lowered below the ambient level. Below 50 ppm, growth of plants is very slow, and if concentrations of CO_2 are maintained below this level for several days, plants become chlorotic and may die.

Rapid growth with high levels of CO_2 requires a high light intensity. If light level is limiting to growth, little or no growth benefit will be obtained from increased CO_2 levels. This response is depicted by Gaastra for sugar beets in Figure 4.4. There is evidence from Wittwer (1970) that corn, a plant requiring high light intensity, shows significant benefit from elevated levels of carbon dioxide only at high light intensities while species with low light requirements respond significantly over a wide range of light intensities.

Certain nutrient deficiencies may be experienced when plants are grown under elevated carbon dioxide levels unless adequate fertilizer is applied to meet the increased demand for nutrients. Direct injury from high levels of CO_2 has not been apparent until concentrations reach 20,000 to 30,000 ppm (2–3%) and that level is maintained for several hours.

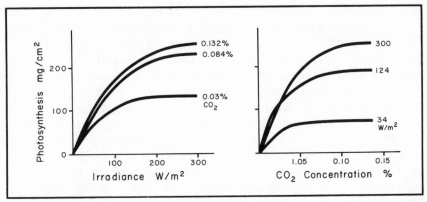

Fig. 4.4. Photosynthetic rates of sugar beet leaves as a function of light intensity and CO_2 concentration. (Gaastra 1959.)

The rapid changes in stomatal opening that occur within a few minutes after variations in the CO_2 level are a major concern in growth chambers. These changes may alter the water status of the leaves and cause rapid changes in the sensitivity of the plant to pollutants or other gasses. Rising CO_2 levels cause rapid closure of stomates, whereas falling CO_2 levels cause the stomata to open. The stomata respond to CO_2 saturation or deficits within the leaf. After closure has been induced by increased CO_2 levels, the stomates will open again when a CO_2 deficit is caused by continued photosynthesis within the leaf. There is evidence that rhythmic opening and closing of the stomata may be induced by an initial rapid increase in CO_2. Thus, as people breathe in controlled environments and cause rapid rises in CO_2 levels, significant fluctuations in stomatal opening can be anticipated.

Sensing or Monitoring Systems

Several systems are available for sensing or monitoring CO_2 levels in growth chambers and each has advantages and disadvantages. Some are inexpensive and simple but limited in their precision, while others are expensive, quite sophisticated in design, and more precise. They also lend themselves to continual automatic monitoring and controlling. Sensors or monitoring systems can be divided into four categories, photochemical sensors, electrochemical sensors, infrared radiation absorption sensors, and other methods (Bailey et al. 1970).

Photochemical

The least expensive and simplest method of measuring CO_2 concentration involves the use of comparative colorimetry (Slavík and Čatský 1965.); however, colorimetry can only provide precision to 50 or 100 ppm and is used for spot measurements rather than for continuous measurement. Sharp (1964) described a method reasonably accurate in the range of greatest interest to the plant physiologist. Gas samples are pumped into a rubber bladder of one-liter capacity and bubbled through an indicator solution of thymol blue. The color of the solution is constantly monitored by means of photoelectric cells, and the CO_2 level is regulated according to the change in color of the indicator solution.

The approximate level of CO_2 in a sample of air can be determined with commercially available chemical gas detectors. The chemical is contained in a glass tube and changes color as an air sample is drawn through the tube. A new tube must be used for each determination. There are other similar systems on the market. Each involves a visual comparison of color changes to determine the level of CO_2. Such manual monitoring systems are available from most laboratory supply houses and greenhouse equipment suppliers.

Electrochemical

Electrochemical or ionic methods involve the measurement of electrical conductivity of the sample in distilled water. A novel type of ionic system for CO_2 measurement and control has been developed in England. A sample of air is bubbled through deionized water, which is passed into a conductivity cell where the resistance is measured electrically. The higher the CO_2 concentration, the lower the resistance. The sensitivity of this system is such that a change in CO_2 concentration of about one part per million produces a change in conductivity of about 3×10^{-9} mho. It provides the precision required for growth chamber research (Fig. 4.5). A small conductivity cell can be placed in each growth chamber, and only two wires need to be run back to the recording equipment. This system makes it possible to have efficient positive monitoring of each growth chamber.

Another electrochemical system described recently (Reyes et al. 1967) for monitoring CO_2 content of blood and other liquids is also suitable for use in sealed environmental systems. The detector configuration incorporates a glass pH electrode and an Ag-AgCl reference electrode joined by an electrolyte solution, with the enzyme carbonic anhydrase as a catalyst. By incorporating a field-effect transistor (FET) in the probe assembly of the detector, the unit has been made relatively stable. It also has a very short response time. These features overcome the earlier limitations of electrochemical sensors.

Infrared Radiation Absorption

The infrared gas analyzer is the most popular instrument for measuring levels of CO_2 in plant growth studies. It permits continuous measurement and recording of CO_2 concentrations with a precision of 1 ppm and even to 0.1 ppm with certain instruments. It works on the principle that the CO_2 molecules absorb energy in the infrared region of the electromagnetic spectrum. Several companies manufacture infrared gas analyzers; their instruments are all very similar. Each has two columns of gas. One column contains a standard zero gas such as nitrogen, which is

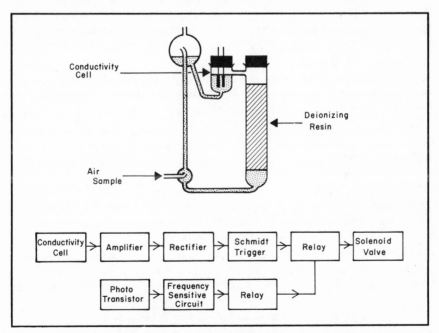

Fig. 4.5. Diagram of a conductivity cell used to monitor and control CO_2 levels in plant growth enclosures. (Bowman 1968.)

free of water and CO_2, while the other contains a sample gas from the growth chamber or other locations. Identical infrared beams pass through the two columns, and the difference in the amount of infrared energy reaching the sensing elements is measured as a deflection of the meter (Fig. 4.6). Some units use a gas-absorbing chamber at the end of the columns while other use radiation-sensing photocells or thermopiles.

About the only variation among infrared monitoring systems is the length of the sample and reference cells. The higher the concentration of the CO_2 to be sampled, the shorter the cell required to obtain the greatest precision. A 0–500 ppm cell is about 9 inches long, a 0–1000 ppm cell is about 5 inches long, and a 0–3000 ppm cell is about 2½ inches long.

For measuring differential CO_2 levels in photosynthesis studies, the basic equipment is modified so that one cell contains the air before it passes over the leaf and the other cell contains the air after it passes over the leaf.

There are many problems associated with infrared gas analyzers that lead to large errors in measurement. (Janac 1970, Brown and Rosenberg 1968). Set-up procedures must be carefully followed to avoid leakage and outgassing contamination from tubing. Cylinders of standard carbon dioxide mixtures obtained from gas suppliers, even those provided with a laboratory analysis, should be calibrated against tanks with known concentrations or by dry chemical determination to insure accurate measurement. Since the concentration of CO_2 in a standard gas cylinder may change after pressure in the tank has fallen below one-third of the original value, recalibration of cylinders during use is necessary.

Users of the infrared analyzer are cautioned against attaching compressed gas cylinders for calibration directly into the tubing to the analyzer. Sudden release of regulator pressure can result in serious damage to the diaphragms in the analyzer. A convenient way to avoid this problem is to insert the outlet tube from the compressed gas cylinder five to seven centimeters into a slightly larger tube leading to the analyzer and suction pump. The flow from the cylinder must be about two times the flow through the analyzer to prevent mixture with room air.

In order to avoid serious breakdown delays, it is advisable to maintain a supply of spare parts for the infrared analyzers.

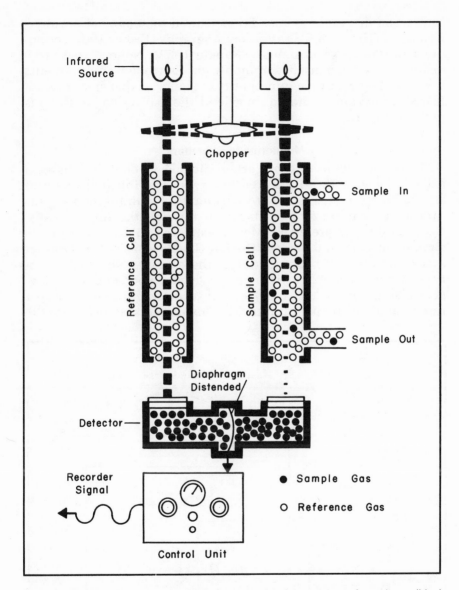

Fig. 4.6. Diagram of a typical infrared analyzer showing the movement of sample gas (black dots) through the sample cell of the analyzer. White dots represent the reference gas. (Beckman 1967.)

Other Methods

Carbon dioxide may also be measured by means of an inter-ferometer (Heys 1967), gas chromatography (Fisher 1962), or liq-uid scintillation spectrometer (Rapkin 1962). None of these pro-cedures has been used commonly in controlled environments because they require equipment that is more expensive than in-frared analyzers or they do not lend themselves to continuous measurement.

Monitoring in Chambers

Effective monitoring of carbon dioxide in chambers is best accomplished with an infrared analyzer. The infrared analyzer should be connected to the growth chamber with tubing that has little or no carbon dioxide diffusion through the tubing walls. Copper tubing is preferred. Most plastic and rubber tubing have significant carbon dioxide diffusion (Lister et al. 1961, Othmer and Frohlich 1955). The outlet of the analyzer should be con-nected to a flowmeter with an accuracy of ±5 percent of the sug-gested flow rate of the analyzer. Most analyzers require a flow rate around 200 ml per minute. The flowmeter is connected to the

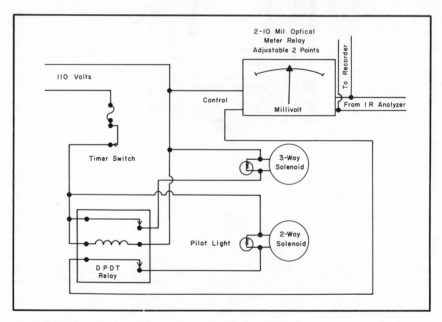

Fig. 4.7. Carbon dioxide control circuit. (Bailey et al. 1970.)

inlet, vacuum side, of a small pump. Diaphragm pumps are often used for they are inexpensive, yet very accurate. The outlet side of the pump is exhausted to the air.

Most manufacturers suggest placing a dryer in the inlet line to the analyzer to remove any water, but if there is a temperature fluctuation (2–3°C), significant absorption and desorption of CO_2 from the desiccant will result, which may produce erroneous CO_2 readings. Desiccants must also be replaced at intervals depending upon the moisture content of the plant-growing atmosphere. Because water vapor response for most analyzers does not exceed 2 percent of the carbon dioxide reading, monitoring of chamber environments is best accomplished without desiccants in the intake line. It is only necessary to place a small glass bottle trap just before the intake of the analyzer to collect any liquid water condensing in the lines and prevent it from reaching the analyzer.

The cost of an infrared analyzer ($2000 to $3000) makes monitoring of carbon dioxide a costly procedure. It is possible to connect more than one chamber into a single analyzer. The number of chambers that can be monitored with one analyzer depends upon the frequency with which the carbon dioxide level is monitored in each chamber. The recommended frequency is at least every 15 minutes so that the influence of human activity around the chambers will be monitored and recorded. The time devoted to each chamber must be long enough to flush out the tubing and sample chambers of the analyzer. A minimum period of one minute is required for this flushing when the tubing to the analyzer has a small diameter and the chamber is within 3 meters of the analyzer. Switching from chamber to chamber can be accomplished through use of a multicam timer with separate gas solenoids placed on the sample lines from each chamber that are activated separately during each cycle. A multiple port sampling unit has been described by Bailey et al. (1970) and Goldsberry (1965). A schematic of the former system is given in Fig. 4.8.

Control Systems

Maintenance of particular carbon dioxide levels is a requirement for many physiological studies, yet it is an almost insurmountable problem for many controlled environment users. The requirement is hard to meet because of the large leakage rate of chambers and the purchase and maintenance costs of adequate monitoring devices. It is easier to maintain a level of carbon dioxide above the ambient level of 350 ppm than it is to maintain a level below ambient levels.

Carbon dioxide levels are raised most efficiently by use of compressed carbon dioxide from gas cylinders. The cylinder outlet should be fitted with a two-stage regulator and a low-pressure regulator. The flow should be controlled from the regulators by low-volume rotometers. A flowmeter with maximum air flow of 100 ml per minute was found effective for reach-in chambers. The flow can be reduced to a very slow rate by use of small diameter (.15 to .25 mm ID) stainless steel capillary tubing and the rate controlled by the length of the capillary tubing.

The simplest procedure for providing CO_2 to the chambers is to leak the gas into the chamber directly from a cylinder with manual adjustment of the flow rate at intervals as needed. Monitor the atmosphere in the chamber continuously or at regular intervals with an infrared analyzer. With this procedure, continual adjustments are necessary as the plants enlarge or as the number of plants is changed. However, it is very difficult to provide effective

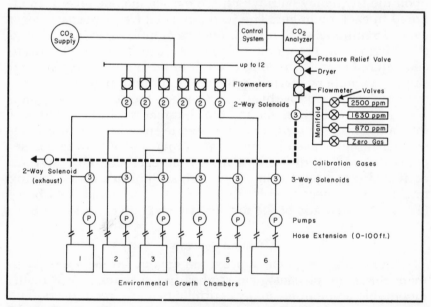

Fig. 4.8. Diagram of pipes and equipment for remote sensing and controlling of CO_2 level in one to twelve growth chambers. The pumps circulate sample gas continuously. (Bailey et al. 1970.)

adjustment for the carbon dioxide fluctuations resulting from people working in or near the chamber.

A controller attached to the infrared analyzer is required to maintain precise chamber levels of CO_2. Various types of recorder

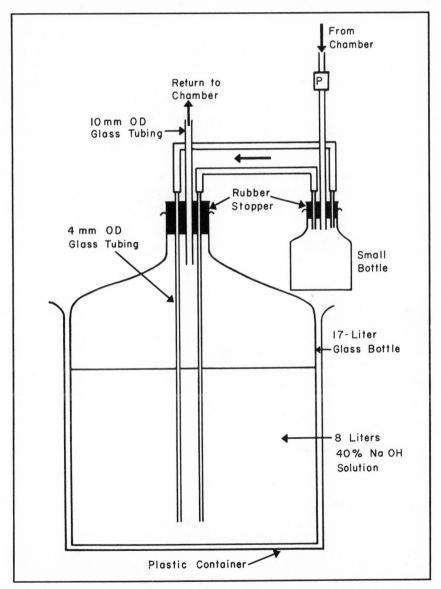

Fig. 4.9. Diagram of carbon dioxide removal system

controllers or electronic controllers cost \$300 or more. Each chamber must have a separate control connected to a solenoid and a cylinder of CO_2. As carbon dioxide decreases in the chamber, a signal, directed from the controller, opens the solenoid and allows the gas to enter the chamber. The flow rate of CO_2 from the cylinder must be low enough so an excess of CO_2 is not added to the chamber during the lag of about one minute that is required for an air sample from the chamber to be pumped back to the analyzer (Fig. 4.8). No more than four chambers can be effectively monitored and controlled with one analyzer, because of the time involved in flushing, sampling, and adding CO_2.

Reduction of carbon dioxide concentrations is accomplished most easily by the addition of fresh air. However, the CO_2 level of fresh air when people are working around the chambers is variable, and control is difficult to attain. It is recommended that sufficient fresh air be provided to have concentrations at the desired level within 15 minutes after people leave the chamber. This recommendation requires the addition of a volume of fresh air equal to the volume of the chamber every 2 or 3 minutes. The small intake fans on chambers cannot provide this quantity of fresh air. The air intake for chambers should draw air from the outside rather than from the chamber area to minimize the CO_2 variation from humans working in the chamber area.

Carbon dioxide in reach-in chambers has been reduced by bubbling a portion of the chamber air through a 40 percent solution of NaOH (Read 1972) and returning this air to the chamber. In this system, when CO_2 concentrations rose above the desired level, the controller on the infrared analyzer was activated and provided a flow of 3 to 4 liters per minute through the NaOH solution. A schematic of the system is given in Figure 4.9. Care must be taken with this system to provide a release to avoid a buildup of pressure in the NaOH container that would break the container and release the concentrated solution.

References

American Chemical Society, Subcommittee on Environmental Improvement, Committee on Chemistry and Public Affairs. 1969. Cleaning our environment: The chemical basis for action. Amer. Chem. Soc., Washington, D.C. 249 pp.

Bailey, W. A.; H. H. Klueter; D. T. Krizek; and N. W. Stuart. 1970. CO_2 systems for growing plants. Proceedings: Controlled Atmospheres for Plant Growth. *Trans. ASAE* 13(2):263–268. Reprinted as ASAE Publ. Proc 270.

Beckman Instruments, Inc. 1967. Beckman continuous infrared analyzers. Bulletin IR-4055-C. Process Instruments Division, Fullerton, Calif. 8 pp.

Bierhuizen, J. F. 1973. Carbon dioxide supply and net photosynthesis. *Acta Hort.* 32:119–126.

Bowman, C. E. 1968. The control of carbon dioxide concentration in plant enclosures. Pages 335–343 in *Functioning of Terrestrial Ecosystems at the Primary Production Level.* Proc. of Copenhagen Symposium. Ed. F. E. Eckhardt. UNESCO, Paris.

Braker, W., and A. L. Mossman, eds. 1971. *Matheson Gas Data Book,* 5th ed. Matheson Gas Products, East Rutherford, N.J.

Brown, K. W., and N. J. Rosenberg. 1968. Errors in sampling and infrared analysis of CO_2 in air and their influence in determination of net photosynthetic rate. *Agron. J.* 60:309–311.

Chapman, H. W.; L. S. Gleason; and W. E. Loomis. 1954. The carbon dioxide content of field air. *Plant Physiol.* 29:500–503.

Fisher Scientific. 1962. The determination of H_2S, SO_2, CO_2 and air using the Fisher gas partitioner. Bulletin 62-6. Catalog 11-126. Applications Instrument Division, Fisher Scientific, Pittsburgh. 2 pp.

Ford, M. A., and G. N. Thorne. 1967. Effect of CO_2 concentration on growth of sugar-beet, barley, kale, and maize. *Ann. Bot.,* N.S. 31:629–644.

Gaastra, P. 1959. Photosynthesis of crop plants as influenced by light, carbon dioxide, temperature, and stomatal diffusion resistance. *Meded. Landb-Hoogesch. Wageningen* 59(13):1–68.

Glatte, H. A., Jr., and B. E. Welch. 1967. Carbon dioxide tolerance: A review. Review 5-67. Environmental Systems Branch, USAF School of Aerospace Medicine, Brooks Air Force Base, Texas.

Goldsberry, K. L. 1965. Multiple port air sampling valve. *Col. Flower Growers Ass. Bull.* 182:3–4.

Heys, G. 1967. Carbon dioxide enrichment in greenhouses. Pages 62–69 in *Proc. Fourth Annual Symposium on Thermal Agriculture.* Natural Gas Processors Association and National Liquid Petroleum Gas Association, Tulsa, Okla.

Holley, W. D. 1965. Carbon dioxide in the greenhouse. Reprint. Minn. State Florists' Bull. (April 1)

Holley, W. D. 1970. CO_2 enrichment for flower production. *Trans. ASAE* 13:257–258.

Janac, J. 1970. The accuracy of the differential measurement of small CO_2 concentration differences with the infrared gas analyzer. *Photosynthetica* 4:302–308.

Kretchman, D. W., and F. S. Howlett. 1970. CO_2 enrichment for vegetable production. *Trans. ASAE* 13:252–256.

Krizek, D. T.; R. H. Zimmerman; H. H. Klueter; and W. A. Bailey. 1971. Growth of crabapple seedlings in controlled environments: Effects of CO_2 level and time and duration of CO_2 treatment. *J. Amer. Soc. Hort. Sci.* 96(3):285–288.

Lemon, E. R. 1970. Mass and energy exchange between plant stands and environment. Pages 199–205 in *Prediction and Measurement of Photosynthetic Productivity.* Proc. Int. Biol. Program, 1969. Trebon, Cz. N.V. Noord-Nederlandse Drukkerije, Meppel, Netherlands.

Lister, G. R.; G. Krotkov; and C. D. Nelson. 1961. A closed circuit apparatus with an infrared CO_2 analyzer and a Geiger tube for continuous measurement of CO_2 exchange in photosynthesis and respiration. *Can. J. Bot.* 39:581–591.

Madsen, E. 1973. The effect of CO_2-concentration on development and dry matter production in young tomato plants. *Acta Agric. Scand.* 23:235–240.

Othmer, D. F., and G. J. Frohlich. 1955. Correlating permeability constants of gases through plastic membranes. *Ind. Eng. Chem.* 47:1034–1040.

Pallas, J. E., Jr. 1970. Theoretical aspects of CO_2 enrichment. *Trans. ASAE* 13:240–245.

Pearman, G. I., and J. R. Garratt. 1973. Carbon dioxide measurements above a wheat crop, 1. Observations of vertical gradients and concentrations. *Agric. Meteor.* 12:13–25.

Rapkin, E. 1962. Measurement of $C^{14}O_2$ by scintillation techniques. Packard Technical Bulletin No. 7. Packard Instrument Company, Downers Grove, Ill. 8 pp.

Read, M. 1972. Growth and tipburn of lettuce: Carbon dioxide enrichment at different light intensity and humidity levels, and rate of incorporation of carbon-14 assimilates into the latex. Ph.D. thesis, University of Wisconsin. 145 pp.

Reyes, R. J.; C. F. Martin; and J. R. Neville. 1967. An improved electrochemical carbon dioxide sensor. SAM-TR 67-54. USAF School of Aerospace Medicine, Aerospace Medical Division, Brooks Air Force Base, Texas. 10 pp.

Roth, E. M. 1964. Carbon dioxide effects. NASA-SP-3006. Nat. Aer. Space Admin. 8 pp.

Šesták, Z.; J. Čatský; and P. G. Jarvis. 1971. *Plant Photosynthetic Production; Manual of Methods*. Dr. W. Junk N. V. Publ., The Hague, Netherlands.

Sharp, R. B. 1964. A simple colorimetric method for the in-situ measurement of carbon dioxide. *J. Agr. Eng. Res.* 9:87–94.

Slavík, B., and J. Čatský. 1965. Colorimetric determination of CO_2 exchange in field and laboratory. *Arid Zone Research* 25:291–298. Methodology of Plant Eco-Physiology. Proceedings of the Montpellier Symposium. UNESCO, Paris.

Wittwer, S. H. 1970. Aspects of CO_2 enrichment for crop production. *Trans. ASAE* 13:249–251.

Wittwer, S. H., and Robb, W. 1964. Carbon dioxide enrichment of greenhouse atmospheres for food crop production. *Econ. Bot.* 18:34–56.

Chapter **5**   **THEODORE W. TIBBITTS**

Air Contaminants

The enclosed environment of growth chambers may be subjected
to air contaminants that do not occur in the open environment of
the field or greenhouse and that are often not considered in evaluating growth chamber research.

Contaminants affecting plants may be generated within the
growth chamber and within the building housing the chamber. It
is generally assumed that plants in a chamber are safe from the
contaminants of the room unless the door is opened. This is not
the case, however, for most commercial chambers exchange
room air with chamber air. Some chambers are specially constructed as sealed units to avoid this problem. The regions of
pressure and suction in separate sections of the circulation system of the chamber may produce rapid air exchange through the
chamber walls. For instance, we found a reach-in commercial
chamber had one complete exchange of room air every 15 minutes even though all the ports were closed.

Most chambers can be subjected to various forms of air contamination, sometimes at a constant level, but more often at in-
frequent intervals and usually from undetermined sources.

Injury to plants is commonly observed in controlled environment chambers. Usually only certain species and sometimes only
certain cultivars of a species will have the observed injury. Of
greatest concern is the presence of contaminants at concentrations that produce no visible injury, but reduce the vigor of the
plants and influence plant response to the experimental treatments. A number of contaminant sources with phytotoxic potential are described in this chapter. Only in limited instances, how-

ever, have specific chemicals been identified and their ability to cause injury to plants in chambers established.

The first months that new chambers are in use are often fraught with problems. The researcher is usually uncertain whether the problems result from his unfamiliarity with the chamber or are caused by contaminants. There have been instances where nearly 6 months of use of a new chamber was required before the problems disappeared.

Interior Chamber Sources

It has been shown that silicone rubber compounds used for sealing between chamber panels released cyclohexylamine that caused injury to plants for 20 months after application to a chamber (Pezet and Gindrat 1978). Injury was most severe upon cucumbers, but tomatoes and beans were also very subject to the injury. The initial symptom of injury on cucumber was downward curling of the leaves followed by marginal and interveinal chlorosis and eventual necrosis. Tomatoes developed similar symptoms and also exhibited distortion of stems and abscission of leaves. French beans developed chlorotic and necrotic spots on the leaves. If concentrations of the cyclohexylamine were high, the tissue of several of the plant species became almost white; if excessive, the plant wilted and died.

Chlorosis of leaves of lettuce plants was observed by the author in chambers for three months after sealing with silicone rubber and other caulking compounds. The chlorosis was most severe on the second, third, and fourth true leaves of the lettuce plants. The injured areas regained chlorophyll when the chlorosis was not too severe.

Volatile emissions from plexiglass are discussed by Rodricks, Cushman, and Stoloff (1967). The use of certain types of plastic shade cloths for reducing light intensity within chambers has been found to be injurious to plants. Bonded plastic shade cloths, not the loose woven cloths, have been found to cause stunting of lettuce plants in studies at the University of Wisconsin. These shade cloths have produced a marginal chlorosis of the first and second trifoliate leaves of bean plants when growth was quite rapid. This injury was similar to that reported by Pezet and Gindrat (1978) and may result from cyclohexylamines released from the bonding compound.

Flexible polyvinyl chloride tubing and butyl rubber tubing

used in chambers were found to be toxic to young cauliflower seedlings by workers at Wellsbourne, England. Rigid polyvinyl chloride and other plastics were not found to produce toxic reactants (R. C. Hardwick, 1972, personal communication). Studies by Sonderen at Wageningen, Netherlands, have demonstrated that certain plasticizers used in the production of flexible polyvinyl chloride tubing have phytotoxic properties. Those plasticizers containing chlorinated hydrocarbons were found to be injurious although the recent studies by Pezet and Gindrat (1978) suggest that cyclohexylamines may be the cause of the problems.

A number of potential contaminant sources exist within the chamber. Fans used for air distribution are a possible source of ozone if electric arcing occurs in the motor. However, no evidence for ozone production or injury in chambers has been reported.

A more common contaminant in chambers is refrigerant leaking from valves or connections within the chamber. The common refrigerants are various compounds of fluorinated hydrocarbons (freon) or salt brines although some use is made of trichloroethylene. No evidence of injury from these compounds has been documented, but it is suspected that trichloroethylene does interact with condensation on plants to cause injury to plants.

The use of commercial steam for humidification may cause problems as a result of chemicals for fungal, bacterial, and pH control in the steam condensate return lines. Cyclohexylamines, in addition to certain hydrazines that have phytotoxic potential, are added to the steam.

Exterior Chamber Sources

The most insidious contaminants causing problems to chamber users are those occurring at irregular intervals due to some unrelated activity in the same building that houses the chamber. By the time symptoms of injury are evident, several days have elapsed since the contaminant production and it is often very difficult to trace back and find the source of the problem.

Examples of this are the use of herbicides such as 2,4-D on an adjacent lawn or the use of pesticides on plants in another chamber or adjacent greenhouses. In Wisconsin, the use of nicotine sulphate bombs for insect control in a greenhouse attached to the growth chamber facility produced sufficient sulphur dioxide to injure alfalfa in reach-in chambers.

Laboratory buildings provide many kinds of atmospheric contamination because of the volatile materials being used in the laboratories. Many researchers have had difficulties in growing plants when chambers were located in laboratory buildings although no specific compounds have been associated with the injuries.

An injury that develops on Solanaceous crops has been shown to be the result of a contaminant in a laboratory building (Mitchell and Vojtik 1967). The contaminant caused lesions and galls on the underside of developing leaves of potatoes, tomatoes, peppers, and eggplant. This contaminant requires at least 30 hours of contact with the plants to produce injury and can be removed by brominated charcoal. Reducing the amount of air circulation in the laboratory building has reduced the amount of injury.

The release of ethylene from ballasts located in the growth chamber area is a potential problem (Wills 1970), yet no evidence has been offered to indicate that ethylene has produced reactions upon plants. Ethylene can also be released by plant material stored in the growth chamber area. The storage of ripe fruit is potentially the most serious hazard.

I have viewed damage to emerging cabbage seedlings at irregular times. The damage has been associated with the replacement of compressors in the growth chamber area, and I suspect that the use of a torch for soldering the connections produced a toxic contaminant, possibly phosgene or a related compound, that injured the plants. The injury upon cabbage results in a graying and shriveling of the cotyledons followed by severe stunting and irregular chlorotic mottling of succeeding leaves.

Materials used in building construction are a large potential source of contaminants. The extensive use of caulking compounds, sealants, paints, and plastic insulation provides volatile compounds for months and even years after construction is completed. Although the injury seen on Solanaceous crops in different facilities may be caused by these emanations, no distinct evidence for phytotoxic effects from these materials has been established.

A discussion of chamber contaminants must emphasize also the problems that can result from outside atmospheric air pollution. The chamber user must always be aware of the hazard of various atmospheric pollutants being drawn into the building or into the chamber and causing injury to plants. In areas subject to

Table 5.1. Air contaminants of controlled environments: Sensitive plants and symptoms of injury

Contaminant or Suspected Source	Sensitive Plants	Symptoms
Cyclohexylamines (silicone rubber sealants) (Pezet and Gindrat 1978)	Cucumbers, Cucumis sativus Tomatoes, Lycopersicum esculentum Lettuce, Lactuca sativa cv. Meikoningen	Downward curling of leaves Marginal and interveinal chlorosis and bleaching of leaves
Polyvinyl chloride (Hardwick 1972, personal communication)	Cauliflower, Brassica oleracea	Not reported
Bonded plastic screening (Tibbitts 1968, personal communication)	Mung beans, Phaseolus angularis	Tip and marginal chlorosis of young leaves
Compressor repairs (Tibbitts 1974, personal experience)	Cabbage, Brassica oleracea	Collapse of cotyledons and mottled chlorosis of youngest leaves
Laboratory buildings (Mitchell and Vojtik 1967)	Solanaceous plants Peppers, Capsicum annum cv. California Wonder Tomato, Lycopersicum esculentum cv. Oxheart Potatoes, Solanum tuberosum cv. Russet Burbank	Lesions on stems and petioles, galls on leaf lamina
Paint (xylene) (Seeley 1976)	Chrysanthemums, Chrysanthemum morifolium cv. Wild Honey Roses, Rosa sp.	No axillary bud growth Leaf abortion and petal spotting
Mercury (Jacobson and Hill 1970)	Roses, Rosa sp.	Petal necrosis
Ethylene (Jacobson and Hill 1970)	Tomato, Lycopersicum esculentum Marigold, Tagetes patula cv. Petite Yellow Orchids, Cattleya sp.	Epinasty of leaves Epinasty of leaves Bud abortion and sepal collapse
Ozone (Jacobson and Hill 1970)	Tobacco, Nicotiana tabacum cv. Bel W-3 Snapbeans, Phaseolus vulgaris cv. Spurt or Tenderette Potatoes, Solanum tuberosum cv. Norland	Flecking on upper leaf surface Flecking on upper leaf surface Dark spots on upper and lower leaf surfaces
Peroxyacetyl nitrate (PAN) (Jacobson and Hill 1970)	Petunia, Petunia hybrida cv. White Cascade Grass, Poa annua Annual Bluegrass	Glazing on under leaf surface Collapse of narrow bands on blades

high levels of photochemical oxidants or sulphur dioxide pollution, the air entering the growth chamber area and the fresh air drawn directly into the chamber should be filtered through sufficient activated charcoal to remove the pollutants.

This chapter ends on a note of caution, for much is yet to be learned concerning the contaminants present in chambers and the amount of injury that is being produced. The growth chamber user does not know whether contaminants are causing subtle injurious effects for he has no plant growth standards for comparison. The development of the base-line growth information by the ASHS Committee on Growth Chamber Environments will provide these needed yardsticks.

References

Jacobson, J. S., and A. C. Hill, eds. 1970. Recognition of air pollution injury to vegetation: A pictoral atlas. Air Pollut. Contr. Ass., Pittsburgh.

Mitchell, J. E., and F. J. Vojtik. 1967. Vein enation and leaf drop of tomatoes grown in controlled environment chambers. *Phytopathology* 57:823 (Abstr.).

Pezet, R., and D. Gindrat. 1978. Injury to plants caused by cyclohexylamine vapors from silicone rubber in growth chambers. *Plant Dis. Reptr.*, 62:101–104.

Rodricks, J. V.; M. Cushman; and L. Stoloff. 1967. Solvent contamination from a volatile component of a fiberglass glove box. *Science* 156 (3782):1648.

Seely, J. G. 1976. Some paints can cause plant injury. *Flor. Rev.* 158 (4103):65, 117–119.

Wills, R. B. H. 1970. Ethylene, a plant hormone from fluorescent lighting. *Nature* 225:199.

Chapter 6

DONALD T. KRIZEK

Air Movement

Wind, or air movement, has long been recognized as an important but complex determinant of plant growth and development. Information on air movement is essential in measuring momentum flux, in determining the extent of heat transfer from the leaves (Mellor, Salisbury, and Raschke 1964, Raschke 1960, Drake 1967, Idso and Baker 1967, Linacre 1967), and in evaluating the transfer of water vapor and carbon dioxide by the aerodynamic approach (Wright and Lemon 1970, Uchijima and Wright 1964).

Numerous studies have been conducted on shelterbelts and wind breaks in the United States and abroad (Geiger 1957, Lorch 1958, Reed and Bartholomew 1930, Warren Wilson 1959, Chang 1968, Waister 1972, Bannister 1976, Van Eimern et al. 1964, Shah 1962). Few experimental studies have been conducted, however, on the influence of air movement on plant growth and development under controlled conditions (Whitehead 1963c, Morse 1963).

This chapter reviews the importance of air velocity, direction of air flow, rate of air exchange, and makeup air and discusses the equipment available for monitoring and controlling air movement in growth chambers.

Parameters of Air Movement in Growth Chambers

Air Velocity

The rate of air movement in a growth chamber influences plant growth by its effect on leaf temperature (Gates 1964, 1968, Cook et al. 1964, Linacre 1967, Matsui and Eguchi 1972), transpiration (Briggs and Shantz 1916, Brown 1910, Firbas 1931,

Kucera 1954, Gaumann and Jagg 1939, Rao 1938, Nakayma and Kadota 1949, Stålfelt 1932, Seybold 1929), evaporation of water from the media and the container, and the amount of carbon dioxide available for photosynthesis (Waggoner et al. 1963, Warren Wilson and Wadsworth 1958; Deneke 1931, Yabuki et al. 1972, Wright and Lemon 1970.) Consequently, leaf size, stem growth and crop yield may be greatly affected by variations in the air velocity.

The primary influence of air movement is on the boundary layer resistance of the leaf surface and hence on the rate at which water vapor leaves the plant as transpiration and enters the air stream. It also influences the rate at which carbon dioxide leaves the air stream and enters the leaf. At wind speeds greater than .89 m sec^{-1} (176 fpm) the boundary layer resistance becomes negligible (Gaastra 1962, Gates 1964, 1968). The importance of air movement in overcoming boundary layer resistance was illustrated by Waggoner, Moss, and Hesketh (1963), who found that photosynthesis was the same under conditions of 200 ppm CO_2 and turbulent air as it was under 300 ppm CO_2 and calm air.

An air velocity of 0.5 sec^{-1}(100 fpm) is generally considered optimum for plant growth under controlled conditions; some researchers, however, recommend rates as high as 1.5 m sec^{-1} (van Bavel 1973). Velocities down to .05 m sec^{-1} (10 fpm) are too low for normal growth (ASHRAE 1977).

A number of studies have been conducted to determine the amount of air movement that a plant can withstand. Most of these have involved wind tunnels in which the plants were subjected to air velocities up to 3.05 m sec^{-1} (600 fpm) in both vertical and horizontal directions. Based on measurements of leaf area and dry weights, the upper limit for air velocity appears to be about .51 m sec^{-1} (100 fpm). For design purposes this may be expressed as a flow through the plant bed of 100 cfm per square foot of growing area (Morse 1963).

Direction of Air Flow

There are three patterns of air flow in a growth chamber—upward, downward, and horizontal. The importance of the direction of air flow depends to a large extent on the size and type of growth chamber, and the type of crop grown.

For large growth chambers, a horizontal pattern of air flow is generally undesirable, since a permanent gradiant in temperature will exist between the intake and the exhaust side of the chamber (Morse 1963). Depending on the radiation load this may be small or quite large (Bailey et al., unpublished results, 1973).

From an engineering point of view, a downward pattern of air movement is considered preferable (Morse 1963); it provides a means of carrying the excess heat away from the lamps and avoids large temperature gradients. Morse and Evans (1962) found little difference, however, in the growth of tomato, lucerne, and subterranean clover plants subjected to vertical movement of air in either direction.

Makeup Air

The need for makeup air will depend to a large extent on whether or not provisions are made for controlling the CO_2 content of the atmosphere. If no provisions are available for CO_2 enrichment, then makeup air is helpful, especially in tightly enclosed growth chambers on growth rooms (Krizek et al. 1970, Bailey et al. 1970).

Where mixed canopies of temperate crops such as wheat are grown with tropical crops such as corn, the CO_2 levels in a growth chamber can drop from 400 ppm to 80 ppm within 30 to 40 minutes after the lights come on (Downs 1975). It is not uncommon to record CO_2 levels as low as 150 to 200 ppm in a chamber containing a full load of actively growing temperate plants such as tobacco (ASHRAE 1977, Krizek et al. 1970, Downs 1975). A fresh air supply is important for providing a minimum supply of CO_2 (Morris et al. 1954) and for avoiding a buildup of ethylene and other volatile substances produced by the plant or released from the growth chamber. The amount of fresh air required to supply CO_2 for photosynthesis has been estimated at a rate of 1.0 to 1.5 cubic meters per minute per square meter of plant growing area (Morse 1963).

Most controlled-environment rooms or growth chambers have ventilation systems that introduce makeup air to prevent CO_2 depletion. In order to maintain near ambient levels of CO_2 in a growth chamber, a large amount of fresh air has to be introduced. Since this is not practical (Downs 1975) CO_2 enhancement of the atmosphere is desirable. The reader is referred to Chapter 4 of this manual for methods of controlling and monitoring carbon dioxide.

Control Systems

In the standard growth chamber little can usually be done to alter the direction or speed of air movement since the ventilation system has been designed into the chamber. In experimental growth chambers (Klueter et al. 1967, Krizek et al. 1968, Bailey et al. 1970) or in specially designed growth chambers (Hammer and Langhans 1972) it may be possible, however, to vary either of these parameters by changing the location of the fans, by adjusting the louvers, or by replacing the fan. In most wind tunnels, a pulley system can be arranged to vary the air velocity.

If several chambers are constructed, air ducts and fan sizes should be identical in each. In setting up an experimental design in the growth chamber, measurements of air velocity need to be taken in advance to determine the existence of gradients within the chamber so that these gradients may be taken into account in assigning plant locations. The importance of louver position in determining growth rate has been shown by Hammer and Langhans (1972).

Most measurements of air movement are taken in an empty chamber for purposes of meeting specifications, but drastic changes in these values may occur once pots are placed in the chamber or the plants are actively growing. The amount of turbulence in a chamber filled with plants that have elongated appreciably will be markedly less than in a chamber with plants at the seedling state or with only a few containers.

Plants should be placed far anough apart to allow for sufficient air circulation throughout the canopy. Strips of paper fastened to a meter stick or other thin piece of wood may be used to detect flutter in the chamber if an anemometer is not available. Many chambers are designed to obtain a slight amount of leaf movement. As a general rule, this means that an air velocity of approximately .3 to .5 m sec^{-1} (60 to 100 feet per minute) is present.

Measurement of Air Flow

Although a wide range of instruments is available for measuring air movement under field conditions (Monteith 1972, Ower and Pankhurst 1966, Slatyer and McIlroy 1961, Wadsworth 1968), many of these instruments are unsuitable for use in growth chambers at low air speeds. There are two basic types of instruments for measuring air movement, cup anemometers and hot-wire anemometers. Both have been used extensively in micrometeorological studies.

Cup anemometers. The rotation of a cup anemometer can be used to operate an electrical contact directly or to interrupt a beam of light falling on a phototransistor (Monteith 1972). There are several phototransistor instruments that incorporate all of the electronics in the housing of the anemometer and have stopping speeds of 5 to 10 cm sec^{-1}. Similar models operate with a mechanical contact and have stopping speeds of 20 to 40 cm sec^{-1}. In both cases the speed of rotation increases nearly linearly with wind speed above the stopping speed.

Hot-wire anemometers. Hot-wire anemometers are used extensively in wind tunnel work and are the type most used in growth chamber measurements. They require a source of constant current or constant voltage and must be kept at a constant temperature. Their advantage for the growth chamber user is that they operate at low windspeeds (as low as a few centimeters per second) and can be used in confined spaces. Their main disadvantages are that they are fragile and have an output that is not linearly related to windspeed (Monteith 1972). They are ideal for measuring air movement near surfaces of the plants and within the canopy. The principle in hot-wire anemometry is that a pure metal wire is heated in an electrical circuit and cooled as it is exposed to the air current. The rate of cooling or change in wire resistance is then used as a measure of air velocity (Caborn 1968). Four materials are commonly used in the construction of hot-wire anemometers: platinum, nickel, tungsten, and platinum-iridium (80:20). Tungsten and platinum-iridium have the highest tensile strength of these four materials and are, therefore, preferred for long wires (Tanner 1963).

Heated thermocouple anemometers. These anemometers are sturdier than hot-wire anemometers and are available commercially. Heated resistance coil and heated thermistor anemometers are also easy to make in the laboratory, but there are few commercial models available.

Pressure tube and vane anemometers. These anemometers are seldom used in growth chamber or micrometeorological studies since their response depends on orientation with respect to wind direction. They work well in wind tunnels and ducts where the direction of air movement is uniform and well defined (Monteith 1972).

Heated thermopile anemometers.   Some air meters incorporate a directly heated thermopile. The Hastings air meter, for
example, has a low voltage a-c bridge circuit with two noble metal
thermocouples as sensing devices. The thermocouples are heated
by alternating current. A change in flow causes a change in temperature of the thermocouples, resulting in a d-c output from the
thermocouples. The thermopile in the head of the probe is fully
temperature compensated for both ambient and rate-of-change of
ambient temperature. Its low mass permits fast response to gusts
and rapidly fluctuating velocities.

A listing of sources for cup anemometers, hot-wire or hot-film
anemometers, thermocouple anemometers, thermistor anemometers, bivane anemometers, and direction vane anemometers is
given in the IBP Manual written by Monteith (1972). The reader
will find a similar list in the mimeographed manual prepared by
Tanner (1963).

There are a number of options available in selecting an anemometer. Some companies manufacture instruments that are
battery operated, but the vast majority operate off of line voltage.
The probe may be unidirectional or omnidirectional. The latter
instrument has been discontinued by certain manufacturers, but
may still be available upon inquiry. A multiple-wire anemometer
is felt to provide the most realistic measure of what the plants
sense, although a unidirectional anemometer is valuable in determining gradients in air movement within the chamber (Downs
1975). Most air meters also have spare or replacement probes,
extension cables with switching devices, and recorder output.

The chief precautions to be taken in using an anemometer are
checking whether the instrument is temperature compensated,
keeping the probe clean, and orienting the probe at the same
angle throughout the chamber. When taking measurements of air
movement in a plant growth chamber or walk-in room, be certain
that the doors are tightly closed and that the investigator is outside the chamber. Since the rate of air movement is influenced by
the density of the crop and the number and proximity of the pots,
it is desirable to obtain measurements at both the beginning and
the end of the experiment.

For a fuller discussion of characteristics of wind-measuring
equipment the reader should examine publications by Middleton
and Spilhaus (1953), Caborn (1967), the British Meteorological

Office (1956), Tanner (1963), Ower and Pankhurst (1966), Monteith (1972), and Kilifarska and Kondacov (1975).

References

ASHRAE. 1977. Chapter 9, Environmental control for animals and plants. Pages 9.1–9.18 in *ASHRAE Handbook of Fundamentals*. Ed. C. W. MacPhee, Am. Soc. Heating, Refrig., and Air Cond. Eng., New York

Bailey, W. A.; H. H. Klueter; D. T. Krizek; and N. W. Stuart. 1970. CO_2 systems for growing plants. *Trans. ASAE* 13(3):263–268.

Bannister, P. 1976. *Introduction to Physiological Plant Ecology*. John Wiley, New York. 273 pp.

Bernbeck, O. 1924. Wind und pflanze. *Flora* 117:293–300.

Briggs, L. J., and H. L. Shantz. 1916. Daily transpiration during the normal growth period and its correlation with the weather. *J. Agr. Res.* 7:155–212.

Brown, M. A. 1910. The influence of air currents on transpiration. *Proc. Iowa Acad. Sci.* 17:13–15.

Caborn, J. M. 1968. The measurement of wind speed and direction in ecological studies. Pages 69–81 in *The Measurement of Environmental Factors in Terrestrial Ecology*. Ed. R. M. Wadsworth. Brit. Ecol. Soc. Sympos. No. 8. 314 pp.

Chang, J. H. 1968. *Climate and Agriculture*. Aldine, Chicago.

Cook, G. D.; J. R. Dixon; and A. C. Leopold. 1964. Transpiration: Its effect on plant leaf temperature. *Science* 144:546–547.

Deneke, H. 1931. Über den Einfluss bewegter Luft auf die Kohlensäureassimilation. *Jb. wiss. Bot.* 74:1–32.

Downs, R. J. 1975. *Controlled Environments for Plant Research*. Columbia University Press, New York. 175 pp.

Drake, B. G. 1967. Heat transfer studies in *Xanthium*. M.S. thesis, Colorado State University, Fort Collins.

Finnell, H. H. 1928. Effect of wind on plant growth. *J. Amer. Soc. Agron.* 20:1206–1210.

Firbas, F. 1931. Die Wirkung des Windes auf die Transpiration. *Ber. dtsch. bot. Ges.* 49:443–452.

Gaastra, P. 1962. Photosynthesis of leaves and field crops. *Netherlands J. Agr. Sci.* 10(5):311–324 (special issue).

Gäumann, E., and O. Jagg. 1939. Der Einfluss des Windes auf die pflanzliche Transpiration II. *Ber. schweiz. bot. Ges.* 49:555–626.

Gates, D. M. 1964. Leaf temperature and transpiration. *Agron. J.* 56:273–277.

——. 1965. Energy, plants, and ecology. *Ecology* 46:1–13.

——. 1968. Transpiration and leaf temperature. *Ann. Rev. Plant Physiol.* 19:211–238.

Geiger, R. 1957. *The Climate Near the Ground*. Harvard University Press, Cambridge, Mass. 494 pp.

Hammer, P. A., and R. W. Langhans. 1972. Experimental design considerations for growth chamber studies. *HortScience* 7:481–483.

Hill, L. 1921. The growth of seedlings in wind. *Proc. Roy. Soc.*, Ser. B. 92:28–31.

Humphries, A. W., and F. J. Roberts. 1965. The effect of wind on plant growth and soil moisture relations: A re-assessment. *New Phytol.* 64:315–318.

Idso, S. B., and D. G. Baker. 1967. Relative importance of reradiation, convection, and transpiration in heat transfer from plants. *Plant Physiol.* 42:631–640.

Kilifarska, M., and I. Kondacov. 1975. A thermocouple battery anemometer for continuous control of low windspeeds in bioclimatic installations. Pages 183–188 in *Phytotronics in Agricultural and Horticultural Research (Phytotronics III).* Ed. P. Chouard and N. de Bilderling. Gauthier-Villars Editeur, Paris.

Klueter, H. H.; W. A. Bailey; H. M. Cathey; and D. T. Krizek. 1967. Development of an experimental growth-chamber system for studying the effects of major environmental factors on plant growth. ASAE Paper No. 67–112. 18 pp.

Krizek, D. T.; W. A. Bailey; and H. H. Klueter. 1970. A "head start" program for bedding plants through controlled environments. Pages 43–53 in *Proc. 3d Nat. Bedding Plant Conf.* Ed. W. H. Carlson. Michigan State University, East Lansing.

——; ——; ——; and H. M. Cathey. 1968. Controlled environments for seedling production. *Proc. Int. Plant Prop. Soc.* 18:273–280.

Kucera, C. L. 1954. Some relationships of evaporation rate to vapor pressure deficit and low wind velocity. *Ecology* 351:71–75.

Linacre, E. T. 1967. Further studies of the heat transfer from a leaf. *Plant Physiol.* 42:651–658.

Lorch, J. 1958. Analysis of windbreak effects. *Bull. Res. Counc. of Israel* 6D:211–220.

Martin, E. V., and F. E. Clements. 1935. Studies of the effect of artificial wind on growth and transpiration in *Helianthus annuus. Plant Physiol.* 10:613–636.

Matsui, T., and H. Eguchi. 1972. Effects of environmental factors on leaf temperature in a temperature-controlled room II: Effects of air movements. *Environ. Control in Biol.* 10:15–18.

Mellor, R. S.; F. B. Salisbury; and K. Raschke. 1964. Leaf temperatures in controlled environments. *Planta* 61:56–72.

Meteorological Office. 1956. *Handbook of Meteorological Instruments.* Part I. H.M.S.O., London.

Middleton, W. E. K., and A. F. Spilhaus. 1953. *Meteorological Instruments.* University of Toronto Press, Toronto. 286 pp.

Monteith, J. L. 1972. *Survey of Instruments for Micro-meteorology.* IBP Handbook 22. Blackwell Scientific Publications, Oxford, England. 264 pp.

Morris, L. G.; J. D. Postlethwaite; and R. I. Edwards. 1954. Ventilation and the supply of carbon dioxide to a glasshouse tomato crop. Tech Memo 87. Nat. Inst. Agr. Eng., Silsoe, England. 14 pp.

Morse, R. N. 1963. Phytotron design criteria-engineering considerations. Pages 20–37 in *Engineering Aspects of Environment Control for Plant Growth.* CSIRO, Melbourne, Australia.

——, and L. T. Evans. 1962. Design and development of CERES: An Australian phytotron. *J. Agr. Eng. Res.* 7(2):128–140.

Nakayma, M., and M. Kadota. 1949. The influence of the wind on the transpiration of some trees II. Bull. Physiogr., Sci. Res. Inst., Tokyo University, Tokyo.

Ower, E., and R. C. Pankhurst. 1966. *The Measurement of Airflow.* Pergamon Press, Oxford, England. 367 pp.

Rao, V. P. 1938. Effect of artificial wind on growth and transpiration in the Italian millet, *Setaria italica. Bull. Torrey Bot. Club* 65:229–232.

Raschke, K. 1960. Heat transfer between the plant and the environment. *Ann. Rev. Plant Physiol.* 11:111–126.

Reed, H. S., and E. T. Bartholomew. 1930. The effects of desiccating winds on citrus trees. Univ. of Calif. Coll. of Agr. Bull. 484. 59 pp.

Seybold, A. 1929. Die pflanzliche Transpiration I. Teil. *Ergebnisse der Biologie* 5:29–165.

Shah, S. R. H. 1962. Studies on wind protection. Inst. for Biol. Field Res., Arnhem. 113 pp.

Slatyer, R. O., and I. C. McIlroy. 1961. *Practical Microclimatology.* UNESCO, Paris.

Stålfelt, M. G. 1932. Der Einfluss des Windes auf die Kutikuläre und Stomatäre Transpiration. *Svensk bot. Tidskr.* 26:45–69.

Tanner, C. B. 1963. Basic instrumentation and measurements for plant environment and micrometeorology. *Soils Bull.* 6. Coll. of Agr., Univ. of Wisconsin, Madison.

Uchijima, Z., and J. L. Wright. 1964. An experimental study of air flow in a corn plant-air layer. *Bull. Nat. Inst. Agr. Sci.* Ser. A 11:19–65.

Van Bavel, C. H. M. 1973. Towards realistic simulation of the natural plant climate. Paper presented at the UNESCO Symposium on Plant Response to Climatic Factors, Uppsala, Sweden.

Van Eimern, J.; R. Karschon, L. A. Razumova, and G. W. Robertson. 1964. Windbreaks and Shelterbelts. World Meteorological Organization Tech. Note 59. 188 pp.

Wadsworth, R. M. 1959. An optimum wind speed for plant growth. *Ann. Bot.* 23(89):195–199.

——, ed. 1968. The Measurement of Environmental Factors in Terrestrial Ecology. Blackwell Scientific Publications, Oxford, England.

Waggoner, P. E.; D. N. Moss; and J. D. Hesketh. 1963. Radiation in the plant environment and photosynthesis. *Agron. J.* 55:36–39.

Waister, P. D. 1972. Wind damage in horticultural crops. *Hort. Abstr.* 42(3):609–615.

Warren Wilson, J. 1959. Notes on wind and its effect in arctic-alpine vegetation. *J. Ecol.* 47:415–427.

——, and R. M. Wadsworth. 1958. The effect of wind speed on assimilation rate: A re-assessment. *Ann. Bot.* 22:285–290.

Whitehead, F. H. 1957. Wind as a factor in plant growth. Pages 84–95 in *Control of the Plant Environment.* Ed. J. P. Hudson. Butterworths, London.

——. 1962. Experimental studies of the effect of wind on plant growth and anatomy II: *Helianthus annuus. New Phytol.* 61:59–62.

——. 1963a. Experimental studies of the effect of wind on plant growth and anatomy III: Soil moisture relations. *New Phytol.* 62:80–85.

——. 1963b. Experimental studies of the effect of wind on plant growth and anatomy IV: Growth substances and adaptive anatomical and morphological changes. *New Phytol.* 62:86–90.

——. 1963c. The effects of exposure on growth and development. Pages 235–245 in *Water Relations of Plants*. Ed. A. J. Rutter and F. H. Whitehead. Blackwell Scientific Publications, London.

——. 1963d. Adapting plants to wind and drought. *Discovery* 24:32–35.

——. 1965. The effect of wind on plant growth and soil moisture relations: A reply to the re-assessment by Humphries and Roberts. *New Phytol.* 64:319–322.

——, and R. Luti. 1962. Experimental studies of the effect of wind on plant growth and anatomy I: *Zea mays*. *New Phytol.* 61:56–58.

Wright, J. L., and E. Lemon. 1970. The exchange of carbon dioxide between the atmosphere and the plant. *Trans. ASAE* 13:238–239.

Yabuki, K.; M. Aoki; and K. Hamotani. 1972. The effect of wind speed on the photosynthesis of rice field. (2). Photosynthesis of rice field in relation to wind speed and solar radiation. Pages 7–9 in *Photosynthesis and Utilization of Solar Energy*. Level III Experiment, JIBP/PP, Japanese National Subcommittee for PP 1971, Tokyo.

Chapter 7

WADE L. BERRY

Nutrition, Containers, and Media

Nutrition, containers, and media are closely related and are discussed simultaneously to present a comprehensive appreciation of this aspect of plant growth in controlled environmental chambers. The necessary environmental conditions that must be provided to the roots for good plant growth are support, water, oxygen, and the essential mineral nutrients. Fortunately, many of the common higher plant species are very similar in their basic requirements, although many differences do exist in their ability to tolerate unfavorable conditions. This chapter concentrates on the different types of growth systems and does not compile a list of anomalies associated with individual species.

Nutrition

Plants are grown in controlled environmental chambers for many reasons, but the common denominator is a desire to grow plants in a predictable way. If plants are to be grown in a predictable way, provision must be made for proper nutrition. Even if a nutritional study is not the objective of growing the plants, nutrition can often influence the final result (Gauch 1972). Deficiencies of one of the essential mineral nutrients and toxicity of a wide variety of substances—including the essential nutrients—can and do occur. Some of the common problems are toxic concentrations of heavy metals, such as zinc and copper from metal parts and the build up of salts from using high salt tap water for replacing water lost to evapotranspiration. The problem is greatest when the root volume has been limited in order to fit many plants in the small area available in most growth chambers.

Many factors affect the absorption and utilization of nutrients, including secondary or tertiary effects of applied treatments. For example, if the root medium temperature drops below 15°C, the availability of phosphate decreases rapidly. If an experiment is run at a low temperature, additional phosphate has to be provided to maintain the plants in a favorable phosphate status. The amount of nutrients required by a plant is also a function of the plant's growth rate. If the growth of the plant is slow due to low temperature, low light, or low carbon dioxide then only small amounts of nutrients are required. If these conditions are exaggerated and the growth rate of the plant is increased correspondingly, then both the amount and frequency of the nutrient additions will have to be increased.

Required Nutrients

Plant physiologists generally consider sixteen elements essential for the normal growth and development of higher plants (Steward 1963). That is, only sixteen elements have been shown to meet the essentiality requirements as set forth by Arnon and Stout (1939). Three of these elements are normally obtained from air and water (carbon, hydrogen, and oxygen). The other thirteen are generally obtained directly from the root medium by way of the plant roots. Six of these elements, the cations, potassium, calcium, magnesium, and the anions, phosphorus, nitrogen, and sulfur, are known as macronutrients because they are found in relatively large concentrations (expressed as small percentages) in plant tissues. The others, iron, chlorine, boron, manganese, zinc, copper, and molybdenum, are known as the micronutrients because their concentration in plant tissues is much lower (measured in parts per million). In addition, sodium has been shown to be essential for halophytes and cobalt is required for the symbiotic fixation of nitrogen by legumes.

Analytical Methods

In many experiments it is desirable to monitor or control the nutrient status of plants or of the root medium. Some of the more common and widely used analytical methods for plant and soil analysis have been compiled by Johnson and Ulrich (1959), Chapman and Pratt (1961), and Black (1965). In the last few years atomic absorption spectroscopy (Christian and Feldman 1970) has become the method of choice for cations and most of the micronutrients. The ammonium molybdate method still seems to be preferred for phosphate (Johnson and Ulrich 1959). The specific ion electrode method for nitrate (Paul and Carlson 1968) is being used increasingly in preference to the phenoldisulfonic acid method or some of the Kjeldahl methods. Chapman (1966), in his book on diagnostic criteria, has brought together most of the information on the use of tissue analysis for assessing the nutrient status of the plant. The use of analytical information for appraising nutritional conditions has been summarized by Walsh and Beaton (1973).

Chemical Units

Equivalents (eq) and parts per million (ppm) are both used to measure nutrient elements. The ppm system is based on weight per unit of substance while the equivalent system is based upon chemical reactivity. Therefore, in chemical reactions, equivalents are the units to use, and when weights or amounts are added, ppm is the more convenient unit.

In plant nutrition, ppm is generally used for the micronutrients where very small amounts are used. Parts per million is a much smaller unit than percentage. One percent is equal to 10,000 ppm.

The equivalent weight of a substance is its atomic or molecular weight divided by its valence or charge. For example, the equivalent weight of potassium (40) is equal to its atomic weight because it has a valence of one. The equivalent weight of calcium is equal to one-half of its atomic weight because it has a valence of two ($40 \div 2 = 20$). When solutions are made on an equivalent basis, they have the same chemical reactivity. That is, solutions with the same number of equivalents have the same number of chemical units to participate in chemical reactions.

A more convenient unit for expressing the concentrations of cations and anions in culture solutions is milliequivalent (meq) which is 0.001 eq. Milliequivalent and ppm are easy to interconvert if the equivalent weight of the substance is known. To convert meq to ppm, multiply by the equivalent weight. To convert ppm to meq, divide by the equivalent weight.

Terminology

Nitrate-nitrogen ($NO_3^- - N$) refers to the amount of nitrogen present in the nitrate form and does not include that present as ammonical-nitrogen (NH_4^+–N). Similarly the amount of sulfur present as sulfate is expressed as $SO_4^=$–S, and phosphorus present as phosphate as $H_2PO_4^-$–P or commonly as just PO_4–P.

Many experiments can use commercial grade chemicals or fertilizers rather than reagent grade chemicals, at a very substantial savings even in small experiments. The percent of the primary nutrients—nitrogen, phosphorus, and potassium—in fertilizers is given by three numbers: as 10=8=6. They are always given in the same order and, in this example, the fertilizer contained 10 percent by weight of total nitrogen, 8 percent available phosphoric acid, and 6 percent water-soluble potassium. The ni-

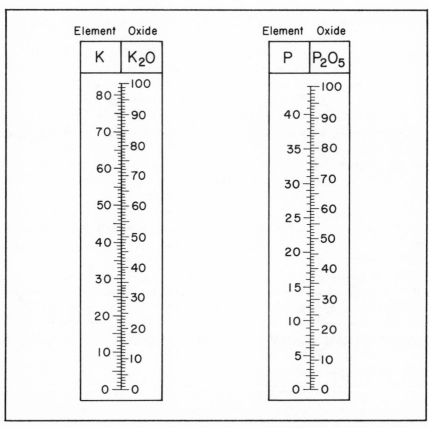

Fig. 7.1. Ratios of potassium and phosphorus oxides to elements. (*Western Fertilizer Handbook* 1965.)

NUTRITION, CONTAINERS, AND MEDIA **121**

Table 7.1. **Conversion of the oxide-elemental forms of phosphorus and potassium**

Conversion to element
$P_2O_5 \times 0.44 = P$
$K_2O \times 0.83 = K$

Conversion to oxide
$P \times 2.29 = P_2O_5$
$K \times 1.20 = K_2O$

trogen is given in elemental terms (N) while the phosphorus and the potassium are given as their oxides (P_2O_5 and K_2O). Table 7.1 and Figure 7.1 give the factors necessary to convert the P and K in commercial fertilizers to elemental amounts.

Containers

The choice of containers is usually determined by custom or availability. Containers of different shapes and materials are available on the commercial market. They are made of wood, clay, and various weights and thicknesses of plastic. It is important that experimenters consider how the choice of container will affect the growth of plants. Container kind and size have a decided influence on the nutritional and water status of the plant.

Container size determines the amount of nutritional reserve available to the plant between nutrient applications. This limitation is most severe for nutrients supplied and retained only in a soluble form, such as nitrogen. The reserve supply of such nutrients is the soil solution. As soon as the nutrients dissolved in the soil solution are absorbed, additional nutrients will have to be added. One way to lengthen the time between nutrient applications is to increase container size and thereby increase the nutrient reserve; the other is to increase the concentration of the applied nutrients. Unfortunately the concentration can be increased only slightly because of the limited salt tolerance of many plants. (Many common plants are ranked according to their salt tolerance in Richards 1954.) But even for the most salt-tolerant plants the gain is relatively small. A trade-off must often be made between small containers and frequent nutrient addition, or large containers and less frequent nutrient additions. A rule of thumb is difficult to establish, but if the roots occupied 50 percent or more of the soil ball by the time the experiment was terminated, the container was probably too small for rapid growth.

In some nutrition experiments, the composition of the container can have a large influence. Clay pots absorb many of the mineral nutrients on their surface and can release them back to the soil at a later time. Therefore, it is very difficult to run deficiency studies on reused clay pots. One way to avoid such contamination is to paint the pot with some inert paint. Alternatively, the pots can be lined with a plastic bag or plastic pots can be used. Because of the dark color of some pots, they can become excessively hot in direct light and the heat may cause root damage. Much of the damage can be avoided if a coat of aluminum paint is applied to the outside of the pot. An aluminum finish will absorb less radiation and thus keep the pots cooler, even in growth chambers.

The composition of the container can also influence the temperature of the rooting media in other ways. As long as the media in a clay pot is wet, water will pass through the porous clay and the entire outer surface of the pot will act as an evaporative cooler. The temperature drop may affect the growth rate of the plant. In direct sun a clay pot will be cooler than a similar plastic pot. The water lost in this way makes water-use calculations very difficult.

Aeration

Good aeration is necessary for the proper growth of most plants. Drainage of the pots must be carefully planned to prevent water-logging and the subsequent development of anaerobic conditions. If the containers do not have drainage holes, it is very difficult to prevent water-logging, because the exact amount of water required by the plant must be added as needed. Also, without drainage, salts will build up because of a lack of leaching. The usual practice is to have drainage holes in the containers, and water until water comes out of the drainage hole. Less water than this will allow a salt build-up in the pot that could eventually reduce growth. Gravel at the bottom of the container does not increase drainage, but only decreases the effective size of the container.

Containers for Nutrient Solutions

A number of different containers have been used with success in solution culture work. Containers of either glass or plastic having a capacity of one liter or greater work well. The size of the container is governed by the same criteria that determines container size when using mixes. The containers should be opaque or painted opaque to prevent the growth of photosynthetic microorganisms (algae). Plastic or masonite pieces painted with valspar varnish works well as a top to hold the plants. Vigorous uniform aeration is necessary to oxygenate and stir the solution and to maintain a uniform composition and temperature. The aeration for all the solutions can conveniently be done from one manifold. A manifold can be constructed from Number 26 hypodermic needles and a large diameter Latex rubber tube. The Latex rubber tube connected to a filtered air supply becomes a manifold for providing a uniform air supply, which is monitored through the hypodermic needles to each individual container by way of a plastic tube. Do not use rubber tubing in direct contact with the nutrient solution because sufficient fixed carbon will leach out to provide a carbon source for proliferation of microorganisms.

Media

Three principal types of root media are commonly used in plant growth chambers: soil mixes, artificial mixes, and solution culture (hydroponics). Each has distinct advantages and disadvantages. No differences in yield and quality of fruit produced should be expected if proper management is used for each type of medium. The choice of root medium, therefore, is primarily dependent on experimental design or specific research objectives and secondarily on cost, degree of care required, availability of materials, or personal preference.

Soils. Soil or soil mixes are not generally recommended for use in growth chambers except as one of the experimental considerations. Natural soil is not a uniform material, and it is difficult to obtain results that are reproducible from experiment to experiment. Natural soil usually contains soil pathogens and weed seeds, is heavy to handle, and is difficult to water in shallow containers.

Artificial mixes. Artificial mixes are commonly composed of sphagnum peat moss, vermiculite, sand, and perlite. These ingredients are used either separately or in combination. This type of medium is generally best suited for the conditions most commonly encountered in growth chambers. Artificial mixes generally have excellent water holding capacity while at the same time they maintain good aeration and light weight. Both peat and vermiculite have a high cation exchange capacity, which gives them a nutrient holding capacity similar to that of soil. These materials in their natural condition are also relatively free of pathogenic microorganisms. The initial composition of artificial mixes is generally more precisely defined than that of natural soil and, therefore, the results of experiments from different times and places can be more readily compared.

Solution culture. Solution culture or hydroponics is the root medium for maximum experimental control, but it is generally not used for growth chamber studies because of the extra work and precautions necessary. In its pure form only water and dilute salts are used for the rooting medium. Plants must be supported by means other than their roots. In commercial practice an inert material such as sand or gravel is often used along with the nutrient solution to enable the roots to provide the necessary physical support for the plants. The use of inert support material does not alter the nutritional aspects of growing plants in culture solution (Hewitt 1966).

The preparation of complete, deficient, or toxic culture solutions requires some degree of proficiency in the sciences of chemistry and botany. Many factors such as concentration, pH, and total amount of nutrient present, are all interdependent and have to be considered simultaneously (Berry and Ulrich 1968). Solution culture offers the experimenter the most control over the growth of the plant. At the same time, the experimenter has to monitor and control more variables to obtain normal plant growth than would be necessary using other systems (Asher et al. 1965).

Preparation of Soil Mixes

In the field most plants will grow on a wide variety of soil types, but only a few soils will produce outstanding yields. Even these soils do not necessarily have the characteristics of a good potting soil. The major purpose in mixing or modifying a soil is to improve its physical characteristics, aeration, and water holding capacity. Incorporating an amendment into the soil dilutes it and may change its texture and structure. Texture refers to the size and distribution of the soil particles (i.e. the percent by weight of sand, silt, and clay). Structure refers to the way the clay, silt, and sand particles are organized into aggregates that behave as units to produce a more or less porous network of particles. The amending agent maintains a separation of the soil aggregation and particles, increasing the number of large pore spaces and making the soil more permeable to water and air. The amendment may also improve the water-holding capacity of the soil. Few amendments improve the physical characteristics of the soil unless they make up more than 30 to 40 percent of the mix by volume.

Inorganic amendments. Sand, when mixed with soil, will help to increase porosity. The sand should be a sharp cement grade, not rounded, and be medium (0.24 to 0.50 mm) to fine (0.10 to 0.25 mm) in grain size. Sands of this size make excellent growing media which, when compacted, still have relatively high infiltration rates. Perlite has many of the same desirable physical characteristics as sand. It increases both the porosity and aeration of the soil in addition to being light weight. Vermiculite has a very high cation exchange capacity, a desirable chemical property that enables the vermiculite to hold the base nutrients and prevent them from being leached out during irrigation, thus adding to the nutrient holding capacity of the mix. Vermiculite is expanded mica; it breaks down with time and loses some of its desirable physical characteristics. In chamber growth use, however, the breakdown of vermiculite is of little consequence during periods shorter than a year.

Organic amendments. Peat moss, ground bark, sawdust, and manure decompose in the soil forming humus, which improves soil structure, thereby increasing infiltration and aeration, and adds to the cation exchange capacity of the soil. The physical characteristics of almost any soil can be improved by the addition of organic material. However, some caution must be used with organic amendments. The decomposition of organic material requires a lot of nitrogen and if about 1.5 percent nitrogen is not present in the added material the organisms breaking it down will take nitrogen from the soil at the expense of the plants. If the material is low in nitrogen, additional nitrogen is needed for the decomposition process to continue without limiting plant growth. Alternatively, slowly broken down organic materials such as peat moss can be used without additional nitrogen, but they do not improve the nutrient value of the mix, although they improve the soil structure and maintain a low pH. Animal manure is a good organic amendment, but frequently contains large quantities of soluble salts that can cause salinity problems if the manure is not composted first.

The acidity of the mix will affect the availability of a number of elements necessary for plant growth. In acid soils (low pH) the amount of available calcium and magnesium is generally very small while iron, manganese, and aluminum are readily available. In very acid soil aluminum and manganese may become toxic. For plants that do best in acid soils, increase the peat moss or ground bark component of the soil mix. Dolomite can be used to reduce the mix's acidity.

Sterilization. Whenever natural soil or a mix containing natural soil is used, the material should be sterilized, either with chemicals or steam. However, when some soils are sterilized with steam, sufficient manganese can be released to become toxic to plants. As a general rule artificial mixes do not require sterilization if care is used in their handling.

Preparation of Artificial Mixes

There are a number of satisfactory artificial soil mixes. Two of the most popular are the University of California and the Cornell peat-lite mixes. The peat-lite mixes are available commercially as Jiffy Mix, Redi-earth, Pro-Mix, and others. The U.C. mix contains various amounts of sand and peat moss. Each sand and peat mix has a special fertilizer mix to go with it for different plants. The U.C. manual (Baker 1957) goes into great detail on how to prepare the U.C. mixes. Boodley and Sheldrake (1963) describe the preparation of the Cornell peat-lite mix.

Peat-lite is a mixture of equal volumes of sphagnum peat and Number Two grade horticultural vermiculite to which have been added fertilizer materials to supply the essential nutrients for plant growth. Peat-lite is an excellent growing medium for general growth chamber use where an artificial mix is desirable to avoid soil diseases or soil textural problems. It is lightweight (about 8 lb ft^{-3}) and has excellent nutrient and moisture-holding characteristics. The chemical fertilizer additives can vary somewhat due to specialized plant requirements, but a good general formula is given in Table 7.2.

Water should be added slowly while mixing the peat and vermiculite, and before adding the fertilizers. This water reduces the loss of small peat particles while mixing and facilitates future wetting in plant containers. Never let peat-lite dry because it is extremely difficult to rewet. The ingredients should be tumbled in a rotary mixer, such as a small concrete mixer, for 10 to 15 minutes to insure a uniform distribution of all particles.

The peat-lite mix contains all the nutrients required for plant growth, but supplemental nitrogen (and possibly phosphorus) will need to be applied in the irrigation water if plants are maintained for a long time on this medium. The amount and frequency will be dependent upon the type, size, and growth rate of the plants.

Table 7.2. **Preparation of peat-lite artificial mix**

To make one cubic yard (0.75 m³) of peat-lite mix:	*Amount*
Sphagnum peat, fluffed up.	16.8 cu ft
No. 2 horticultural grade vermiculite	16.8 cu ft
Water	12.5 gal (47 L)
Dolomite lime	5.6 kg
4-10-10 fertilizer	3.2 kg
138 Fe iron chelate	35 g
or	
330 Fe iron chelate	21 g
Micronutrient concentrate	236 ml
To make one liter of micronutrient concentrate:	*Amount*
Copper sulfate (CuSO₄ · 5H₂O)	14 g
Zinc sulfate (ZnSO₄ · 7H₂O)	6 g
Manganese sulfate (MnSO₄ · H₂O)	4 g
Boric acid (H₃BO₃)	14 g
Add water to make one liter.	

Note: One cu ft of baled peat equals approximately 1.5 cu ft loose peat. A reduction in volume follows mixing since dissimilar particles are being mixed. Peat resists wetting; add water slowly to provide a thorough initial wetting.

Preparation of Nutrient Solutions

In solution culture the required mineral nutrients are supplied as a dilute solution of their salts, and oxygen availability is maintained by constant aeration of the solution. The preparation of the nutrient solution is relatively simple if the supply water is pure (distilled water) and pure salts are used (reagent grade). Otherwise, the actual composition of the materials used will have to be considered and in some cases adjustments made. For example, in some areas tap water contains sufficient levels of boron to be toxic to sensitive plants (concentrations greater than 0.5 ppm). The preparation of nutrient solutions should be done under conditions of extreme cleanliness. When plants are grown in solution culture, there is little buffering capacity in the system and very small amounts of biotoxic materials will result in growth reduction (Berry 1977). For instance, if copper, such as copper tubing, comes in contact with the nutrient solution, sufficient copper will be dissolved to cause acute copper toxicity to the plants growing in that solution; but in a soil medium only the roots in immediate contact with the copper would be killed. Assuming that pure water and salts are available, the following solution (Tables 7.3, 7.4, and 7.5), developed for tomatoes, has been found to satisfy the nutrient requirements of many types of plants. Slight modifications of this solution may prove somewhat better depending upon the plant species being grown and the type of response desired; that is, whether top, root, fruit, flower, or some other aspect of growth or yield is being encouraged. If corn and certain other monocots are grown in this nutrient solution, the amount of added iron will have to be increased 2 to 4 times. Researchers in plant nutrition will recognize the formula presented in Table 7.3 as a minor modification of half-strength Hoagland's solution (Hoagland and Arnon 1950). The stock solutions can be stored successfully, and for convenience can be made up in batches large enough to supply requirements for two or three months. It will be necessary to prepare the two concentrates (1 and 2) in separate containers in order to avoid a precipitate of calcium and iron

Table 7.3. **Preparation of stock concentrates**

Stock concentrate 1	Amount per liter
Potassium nitrate (KNO_3)	50.55 g
Potassium di-hydrogen phosphate (KH_2PO_4)	27.22 g
Magnesium sulfate ($MgSO_4 \cdot 7H_2O$)	49.30 g
Sodium chloride (NaCl)	5.85 g
Micronutrient concentrate	100 ml

Add water to make one liter and mix thoroughly to dissolve all salts.

Micronutrient concentrate	Grams per liter
Boric acid (H_3BO_3)	2.85
Manganese sulfate ($MnSO_4 \cdot H_2O$)	1.538
Zinc sulfate ($ZnSO_4 \cdot 7H_2O$)	0.219
Copper sulfate ($CuSO_4 \cdot 5H_2O$)	0.078
Molybdic acid (MoO_3) (85%)	0.020

Add water to make one liter and mix thoroughly to dissolve all salts.

Stock concentrate 2	Grams per liter
Calcium nitrate ($CaNO_3$) $\cdot 4H_2O$*	118.08
Sequestrene 330 Fe†	5.0

Make a slurry of the iron chelate in a small amount of water before adding to the calcium nitrate concentrate.

*If commercial calcium nitrate (Norsk Hydro) is used it has a formula of $5\ Ca(NO_3)_2\ NH_4NO_3 \cdot 10\ H_2O$; add only 88.8 g/L to provide the right amount of nitrogen. The calcium content will be somewhat lower.

†Other iron chelates will work (see Jacobson 1951): Sequestrene 138 Fe at 8.3 g/L or NaFe EDTA at 4.2 g/L.

phosphates. Fiberglass or plastic-lined tanks are recommended. The tanks should also be opaque because this solution is excellent for the growth of green algae in the presence of light.

Table 7.4. **Approximate concentration of nutrients in final solution (200:1 dilution of stock concentrates)**

	NO_3^--N	$H_2PO_4^--P$	K^+	Na^+	Ca^{++}	Mg^{++}	$SO_4^=-S$
Atomic weight	14.01	30.97	39.10	22.99	40.08	24.31	32.06
ppm	105.0	31.0	140.0	11.0	100.0	24.0	32.0
meq/L	7.5	1.0	3.5	0.5	5.0	2.0	2.0
	Cl^-	Fe^{++}	B^{+++}	Mn^{++}	Zn^{++}	Cu^{++}	MoO_4-Mo
Atomic weight	35.45	55.85	10.81	54.94	65.37	63.54	95.94
ppm	18.0	2.5	0.25	0.25	0.025	0.01	0.005
meq/L	0.5	0.089	0.069	0.009	0.00076	0.00031	0.00010

Mixing the stock solutions. The dilute nutrient solution used for plants is prepared by adding equal quantities of the two stock concentrates to an appropriate container or reservoir and diluting to the volume required for a 200:1 dilution. To make 10 liters of nutrient solution, fill an appropriate container three-quarters full of water, add 50 ml of each stock solution, and make to volume. More concentrated stock solutions than those described should not be attempted since precipitates will form, removing some of the nutrients from solution.

This solution is not only useful in solution culture, but it can also be used as a dilute fertilizer for both soil and artificial mixes. It is a complete fertilizer that contains all the essential mineral nutrients. When using this dilute solution for nutrient application, frequent applications are required. Tomatoes grown in straight Number Two grade horticultural vermiculite (two plants per 5-gallon container) do very well when watered twice a day with this nutrient solution. However, it is important that sufficient water be added each time for drainage in order to prevent a build-up of salt.

Adding makeup solution. Tomatoes and cucumbers can be grown in this solution on a long-term maintenance basis without the necessity of routinely dumping or renewing the solution. The strength of the nutrient solution can be maintained at a favorable level by monitoring the total amount of salt, by daily mea-

Table 7.5. **Conversion factors for modifying nutrient solutions**

		Molecular weight	Atomic weight	Grams to make		meq to ppm	ppm to meq	ppm in one g/L
				100 meq/L	1 ppm			
Potassium nitrate	KNO₃	101.11						
	K⁺		39.10	10.11	2.59	0.025	39.10	0.39
	NO₃⁻–N		14.01	10.11	7.22	0.071	14.01	0.14
Potassium di-hydrogen phosphate	KH₂PO₄	136.09						
	K⁺		39.10	13.61	3.48	0.025	39.10	0.29
	H₂PO₄⁻–P		30.97	13.61	4.39	0.032	30.97	0.23
Magnesium sulfate	MgSO₄ · 7H₂O	246.48						
	Mg⁺⁺		24.31	12.3	10.14	0.082	12.16	0.10
	SO₄⁼–S		32.06	12.3	7.69	0.062	16.03	0.13
Sodium chloride	NaCl	90.44						
	Na⁺		22.99	9.04	3.93	0.043	22.99	0.25
	Cl⁻		35.45	9.04	2.55	0.028	35.45	0.39
Calcium nitrate	Ca(NO₃)·4H₂O	236.15						
	Ca⁺⁺		40.08	11.81	5.89	0.050	20.04	0.17
	NO₃⁻–N		14.01	11.81	16.86	0.071	14.01	0.06
Boric acid	H₃BO₃	61.83						
	B		10.81		5.72			0.18
Manganese sulfate	MnSO₄ · H₂O	169.01						
	Mn⁺⁺		54.94	8.45	3.08	0.036	27.47	0.33
Zinc sulfate	ZnSO₄ · 7H₂O	287.54						
	Zn⁺⁺		65.37	14.38	4.40	0.031	32.69	0.23
Copper sulfate	CuSO₄ · 5H₂O	249.68						
	Cu⁺⁺		63.54	12.48	3.93	0.031	31.77	0.25
Molybdic acid	MoO₃ (85%)	143.94						
	Mo		95.94		1.50			0.67

surements of the electrical conductivity. Whenever the conductivity drops below the reference level, the conductivity of the original nutrient solution, equal amounts of both the concentrates are added until the reference level is restored. Under favorable conditions the reference level for this nutrient solution is approximately 1 millimhos/cm. In an average situation about 5 ml of each concentrate will be required per liter of makeup water to maintain the reference conductivity. However, this amount can vary slightly depending on climate and other environmental conditions. When the plants are growing fast (high light) they will require more nutrients than when the conditions are less favorable (low light).

Electrical conductivity is not a direct measure of the nutrient status of the solution. Therefore, if trouble occurs, a complete chemical analysis of the nutrient solution along with visual symptomology (Sprague 1964) may provide clues for diagnosing possible deficiency or toxicities. If the supply water is not pure, dissolved salts in excess of the plant's needs or those not required by the plant will raise the conductivity reading of the nutrient solution without adding nutrient value. A nutrient deficiency problem could then occur if the conductivity is not adjusted to account for these salts, even when they are excesses of some of the essential nutrients, calcium, magnesium, or sulfate. When water softeners are used to replace the other cations with sodium even more salt damage will result. In fact it is recommended that softened water never be used for horticultural purposes.

Controlling pH. In solution culture, pH is a variable that can easily be monitored and that provides much information about the nutrient solution. It can often indicate an imbalance in nutrient absorption or contamination by foreign material. If the plant absorbs more anions than cations, the solution will become basic and if it absorbs more cations than anions the solution will become acidic. A change in pH may affect the solubility and uptake of nutrients and therefore may greatly alter plant growth. The nutrient solution prepared from Table 7.3 has a pH of about 6 and will remain fairly constant when tomatoes are grown in it.

The pH of the solution can be controlled in a number of ways. For minor adjustments, increasing concentration and, therefore, the absorption, of K^+ relative to the other cations will cause an acidic shift in the solution. Increasing the relative concentration of NO_3^- will cause a basic shift. Larger adjustments can be made with small amounts of dilute sodium hydroxide (NaOH) or hydrochloric acid (HCl). If it is desirable to maintain the solution at a constant pH, less soluble salts such as calcium sulfate ($CaSO_4$), calcium phosphate ($Ca_3(PO_4)_2$) or calcium carbonate ($CaCO_3$) can be used. A pH of 4.0 can be obtained by adding excess $CaSO_4$ and a pH of 7.0 by adding excess $CaCO_3$. Intermediate pH's can be obtained by using mixtures of the above three calcium salts.

The nutrients in this solution are available over a wide pH range. However, under alkaline conditions, phosphate tends to precipitate slowly. This action can be corrected by lowering the pH or routinely adding additional phosphate.

Table 7.6. **Composition of concentrated reagent acids and hydride solutions**

	HCl	HNO₃	HClO₄	H₂SO₄
Molecular wt.	32.46	63.01	100.46	98.08
Specific gravity of conc. reagent*	1.19	1.42	1.67	1.84
Strength (w/w) of conc. reagent (%)*	37.2	70.4	70.5	96.0
Molarity of conc. reagent	12.1	15.9	11.7	18.0
ml conc. reagent per liter for a one molar solution	82.5	63.0	85.5	55.5

	H₃PO₄	NH₄OH	NaOH	KOH
Molecular wt.	98.00	35.05	39.99	56.10
Specific gravity of conc. reagent*	1.70	0.90	1.54	1.46
Strength (w/w) of conc. reagent (%)*	85.5	57.6	50.5	45.0
Molarity of conc. reagent	14.8	14.8	19.4	11.7
ml conc. reagent per liter for a one molar solution	67.5	67.5	51.4	85.5

*Approximate.

Micronutrient variations. The micronutrient concentration in the nutrient solution can rise to excessive levels from impurities found in the system. Significant amounts of micronutrients are at times found in the macronutrient salts and the supply water, or leached from the plumbing and containers of the system. Any brass or copper parts in contact with the culture solution may contribute toxic amounts of zinc and copper to the solution. Gravel and sand used for support may contribute impurities. Choose only inert material such as quartz sand and wash it thoroughly before using. For deficiency studies an additional wash in 1.2M HCl followed by a number of distilled water rinses is recommended.

Nutrient solution deficiencies. In many experiments plants are grown in solutions deficient in one of the essential elements, in order to demonstrate or study unusual deficiency symptoms, or to study the effect of nutrient stress upon other growth factors. Tables 7.7 and 7.8 give formulas for making solutions deficient in each of the macronutrients. If the stress or deficiency symptoms are desired on larger or more mature plants, it is best to grow the plants for a while in a complete nutrient solution before changing to a deficient solution. Seeds germinated on truly deficient nutrient solutions will not grow beyond the seedling stage thus only very small plants showing very acute deficiency symptoms will be obtained.

Plant preparation. Germination of seeds for liquid culture solutions is not generally done directly in the culture solution because of mechanical problems in holding the seed. Also if seeds are germinated elsewhere, uniform seedlings can be selected to start the experiment. For most experimental purposes the seeds can be germinated in trays of vermiculite. The trays should be deep enough so that the roots will not hit bottom before being transplanted to the culture solution. Water the germinating seeds with one-fifth strength nutrient solution. Most plants can be transplanted as soon as the first true leaf is partially emerged. The earlier the plant is transplanted the less severe the shock, but at the same time small seedlings are very tender and more subject to physical damage. The seedlings are removed for transplanting by carefully submerging the tray with the seedlings in water and gently floating the seedlings from the vermiculite. If small particles adhere to the root, do not attempt to remove them for the roots are tender and a few particles will not interfere with the experiment, except in nutrient studies. Carefully wrap a wad of Dacron, plastic foam, or nonabsorbent cotton around the stem of the seedling before supporting it over the one-half strength nutrient solution, to allow the stem to expand and grow. At this time it is extremely important to make sure that the nutrient solution covers the entire root of the seedling and that the root does not hang up on the side of the container, but does in fact dangle down into the nutrient solution.

Table 7.7. **Composition of deficient nutrient solutions**

	Complete nutrient solution (meq/L)	Deficient treatments (meq/L)					
		$-N$	$-P$	$-K$	$-Ca$	$-Mg$	$-SO_4$
Cation							
Na	0.5	0.5	0.5	0.5	0.5	0.5	0.5
K	4	4	4	0	8	4	4
Mg	2	2	2	2	3	0	2
Ca	5	5	5	9	0	7	5
Anion							
NO_3	7	0	7	8	7	7	7
PO_4	1	2	0	1	1	1	2
SO_4	3	5	4	2	3	3	0
Cl	0.5	4.5	0.5	0.5	0.5	0.5	2.5

Note: Micronutrients in mg/liter are 0.25 B, 0.25 Mn, 0.025 Zn, 0.01 Cu, 0.005 Mo, and 2.5 Fe as the EDTA complex. All solutions are designed to have the same total salt concentration.

Table 7.8. **Preparation of nutrient solutions deficient in one of the essential macronutrients**

Stock solution		Complete nutrient solution (ml/L)	Deficient treatments (ml/L)						Molecular weight
			−N	−P	−K	−Ca	−Mg	−SO$_4$	
1M	Ca(NO$_3$)$_2$ · 4H$_2$O	2.5	0	2.5	4	0	3.5	2.5	236.16
1M	KNO$_3$	2	0	2	0	7	0	2	101.1
0.5M	K$_2$SO$_4$	1	3	2	0	0	3	0	174.26
1M	MgSO$_4$ · 7H$_2$O	1	1	1	1	1.5	0	0	120.39
1M	KH$_2$PO$_4$	1	1	0	0	1	1	2	136.09
1M	NaCl	0.5	0.5	0.5	0.5	0.5	0.5	0.5	58.45
	Sequestrene 330 Fe concentrate* 50 g/L	0.5	0.5	0.5	0.5	0.5	0.5	0.5	
	Micronutrient concentrate (Table 7.3)	0.5	0.5	0.5	0.5	0.5	0.5	0.5	
0.05M	Ca(H$_2$PO$_4$)$_2$	0	10	0	10	0	0	0	170.07
1M	CaCl$_2$	0	2	0	0	0	0	0	110.99
1M	MgCl$_2$	0	0	0	0	0	0	1	95.23

Note: Micronutrients in mg/liter in the final solution are 0.25 B, 0.25 Mn, 0.025 Zn, 0.01 Cu, 0.005 Mo, and 2.5 Fe as the EDTA complex. All solutions are designed to have the same total salt concentration.

*Store in refrigerator.

References

Arnon, D. I., and P. R. Stout. 1939. The essentiality of certain elements in minute quantity for plants with special reference to copper. *Plant Physiol.* 14:371–375.

Asher, C. J.; P. J. Ozanne; and J. F. Lonergan. 1965. A method for controlling the ionic environment of plant roots. *Soil Sci.* 100:149–156.

Baker, K. F. 1957. *The U.C. System for Producing Healthy Container-Grown Plants.* Univ. of Calif. Agri. Sta. Manual 23. 332 pp.

Berry, W. L. 1977. Dose-response curves for lettuce subjected to acute toxic levels of copper and zinc. In *15th Annual Handford Life Science Symposium.* Ed. H. Drucker and R. E. Wilding. ERDA Symposium Series Conf. 750929.

Berry, W. L., and A. Ulrich. 1968. Cation absorption from culture solution by sugar beets. *Soil Sci.* 106:303–308.

Black, C. A., ed. 1965. Methods of soil analysis. Agronomy Series 9:771–1572. Amer. Soc. Agron., Madison, Wis.

Boodley, J. W., and R. W. Sheldrake. 1963. Artificial soils for commercial plant growing. New York State College of Agri. Cornell Univ. Ag. Ext. Bull. 1104.

Chapman, H. D., ed. 1966. Diagnostic criteria for plants and soils. Univ. of Calif. Agr. Pub., Riverside.

Chapman, H. D., and P. F. Pratt. 1961. Methods of analysis for soils, plants and waters. Univ. of Calif., Div. of Agr. Sci., Riverside.

Christian, G. D., and F. J. Feldman. 1970. Atomic Absorption Spectroscopy: Applications in Agriculture, Biology, and Medicine. Wiley-Interscience, New York. 490 pp.

Gauch, C. H. 1972. *Inorganic Plant Nutrition.* Dowden, Hutchinson and Ross, Stroudsburg, Penn.

Hewitt, E. J. 1966. *Sand and Water Culture Methods Used in the Study of Plant Nutrition.* Tech. Comm. No. 22. 2d ed. Commonwealth Bureau of Hort. and Plantation Crops, East Melling, Maidstone, Kent, England. 547 pp.

Hoagland, D. R., and D. I. Arnon. 1950. The water-culture method for growing plants without soil. Calif. Ag. Exp. Sta. Cir. 337.

Jacobson, L. 1951. Maintenance of iron supply in nutrient solutions by a single addition of ferric potassium ethylenediamine tetra-acetate. *Plant Physiol.* 26:411–413.

Johnson, C. M., and A. Ulrich. 1959. *Analytical Methods for Use in Plant Analysis.* Univ. of Calif., Div. Agri. Sci. 766 pp.

Paul, J. L., and R. M. Carlson. 1968. Nitrate determination in plant extracts by the nitrate electrode. *J. Agr. Food Chem.* 16:766–768.

Richards, L. A., ed. 1954. Diagnosis and improvement of saline and alkali soils. USDA Ag. Hb. 60.

Shaw, Earle J., ed. 1965. *Western Fertilizer Handbook.* California Fertilizer Association. Soil Improvement Committee. 200 pp.

Sprague, H. B., ed. 1964. *Hunger Signs in Crops.* David McKay, New York.

Steward, F. C., ed. 1963. *Plant Physiology, A Treatise.* Vol. III. Academic Press, New York.

Walsh, L. M., and J. D. Beaton, eds. 1973. Soil testing and plant analysis. Soil Sci. Soc. of Amer., Madison, Wisc.

Chapter **8**

HERSCHEL H. KLUETER, WILLIAM A. BAILEY, WILLIAM A. DUNGEY, DONALD T. KRIZEK, AND GEOFFREY BURDGE

Watering Systems

Plant growth under controlled environments requires a carefully regulated supply of water and nutrients. These requirements can be supplied by mist irrigation, drip irrigation, or subirrigation. The value of an automatic supply of water or nutrients, or both, has long been recognized by the growers of container plants in greenhouses, as well as by nurserymen. Several experimental systems have been designed for laboratory use during the past twenty-five years. In setting up an automatic system, each investigator must consider several questions. How often should water be supplied? How long should the watering regime last? How should the water be applied? Should nutrient solution be applied during the entire regime? What are the required characteristics of the soil? What types of containers will be used? What are the specific needs of the plants? What type of growth is desired?

Most persons would agree that an "ideal" system would include a regime of water and nutrients that would maximize the desired effect, plant growth. As the growth rate of a plant varies with time and stage of development, its requirements for moisture and nutrients are not constant. Flexibility is an essential design criterion for an automatic watering and nutrient system. This chapter briefly reviews the historical development of automatic watering and nutrient systems, describes the components of good systems, gives some of the advantages and disadvantages of automatic systems, and lists some of the special precautions needed.

136

Early Development

One of the first watering schemes was developed by Livingston (1908). Plants in cylindrical cups were placed on a moist bed, which served as the source of water. Later, double-wall irrigator pots were developed by Wilson (1929), but they were not satisfactory.

Richards and Loomis (1942) developed a system called an autoirrigator, "a single piece construction having a glazed outer wall 1/4 inch thick, a porous inner wall 1/4 inch thick, and an inter-wall space of 1/4 inch for the water supply." It provided constant soil moisture and an easy method for measuring the average moisture content. Its usefulness was limited because of the slow rate at which surrounding soil was saturated. These authors were among the first to advocate the use of an automatic watering system in seed germination, early seedling growth, rooting of cuttings, and the study of the development and control of soil-borne insects and pathogens.

De Vries (1963) designed a watering system in which a constant level of water was maintained in a bed of sand. The plants received water through capillary action; but because of the constant water level, leaching could not be achieved. Fibrous mats have also been used. Larson and Hilliard (1976) found that these mats worked well on poinsettias. Their application in growth chamber studies has not yet been evaluated.

The drip, or trickle, watering and nutrient system has had many proponents. Bean and Wells (1957) found that use of plastic capillary tubes was the most efficient technique for applying the drip method to the watering of plants. Marcussen and Scott (1965) found that one of the drawbacks of the drip system is that an even water pressure must be maintained. The misting system is perhaps the most common type of automatic system, but its application has been largely confined to greenhouses (Aljibury 1966, Bean et al. 1957, Geiger 1960, Nakasone and Bowers 1959, and Vanstone 1959). This system consists of several equally-spaced nozzles that emit a fine mist when water or nutrient is supplied to potted or bedded plants or to cuttings.

Robothaln and Pickett (1972) developed a nutrient system with automatic programmable controls for phytotrons. The systems contain solid-state electronics for control of the type of nutrient solution applied in various experiments and for the time, rate, and frequency of its application. These systems work well for large operations, but at present they are too costly for small experiments.

Automatic Watering and Nutrient Systems

An effective automatic watering and nutrient system has four basic requirements: a source of clean water and water-soluble fertilizer, sufficient pressure throughout the system, uniform distribution throughout the growing area, and automatic control of duration and frequency of application.

Supply of Water and Nutrients

Tap water is generally not adequate for growth chamber studies in which quality of the water supply is critical. Demineralized or distilled water is often necessary. Several techniques have been developed for provision of water and nutrient solution. In the simplest method one tank is used for both batch mixing and nutrient supply. A more satisfactory method employs two tanks (Fig. 8.1A), one above the other. The upper tank is used for batch mixing, and the lower one is used as the supply.

The control system consists of a float valve, a three-way solenoid valve, a double-pole, double-throw (DPDT) relay, a (normally closed) float switch, and a manual momentary contact switch for convenient access. The three-way solenoid valve is normally de-energized so the upper tank will drain into the lower one by gravity. The float valve in the lower tank maintains the desired level, until the upper tank is empty. When the start button (Fig. 8.1B) is pushed, the three-way solenoid (two-way solenoids could also be used) is activated; it allows water from the supply line to enter the upper tank and simultaneously closes the drain to the lower tank. The relay locks the solenoid into the activated position until the upper tank is filled. A measured amount of water-soluble fertilizer is added during tank filling. When the tank is filled, the float switch at the top of the upper tank disengages the relay and the solenoid, thereby allowing the nutrient solution to drain into the lower tank upon demand.

Commercially available proportioning injectors can be used rather than large supply tanks. As water flows through an injector, concentrated nutrient solution is automatically added. Standard ratios of 1:100 to 1:200 are used. The diluted nutrient solution should be checked periodically for accuracy. If the flow

Table 8.1. **Materials used for a system of automatic watering**

Description	Model no.	Manufacturer
Supply system		
Submersible pump,	5' head	Little Giant Pump Co.
1/40-hp, 205 gph	1P-372	3810-A-N Tulsa St.
1/15-hp, 337 gph	1P-373	Oklahoma City, OK 73112
1/3-hp pump with insulated, glass-lined, 22-gal. tank with pressure control		Montgomery Ward 1000 S. Monroe Street Baltimore, MD 21232
Distribution system		
36 on-off tubes, 5/64"ID, 1/8"OD, 36" long	NF 36	Chapin Watermatics, Inc. 368 N. Colorado Avenue Watertown, NY 13601
36 metal grommets, 1/8" ID 1/4" long		Chapin Watermatics, Inc.
25' roll black flexible tubing, 1/4" ID, 3/8" OD, 1/6" wall	Tygon R-3400	Norton Plastics and Synthetics Div. Akron, OH 44309
6 mist nozzles	Floral mist	Any florist or greenhouse supply company
Control system		
24-hr clock with 15-min increments	8001	Tork Time Controls, Inc. 100 Grove St. Mt. Vernon, NY 11551
60-min timer with 1-min increments, 120 volt	60M8001	Tork Time Controls, Inc.
30-min timer with 30-sec increments, 120 volt	30M8001	Tork Time Controls, Inc.
12-min timer with 12-sec increments	12M8001	Tork Time Controls, Inc.
5/6"-orifice solenoid, motor operated, spring closure	700AH-1	Penn Controls 2221 Camden Court Oak Brook, IL 60521
Stepping switch, 120V, 60 Hz coil	Sol-o-Step 21 point	Guardian Electric Manufacturing Co. 1550-T W. Carroll Chicago, IL 60521
2- to 120-sec variable reset timer	305-C	Automatic Timing & Control 203 S. Gulph Rd. King of Prussia, PA 19406
4-pole, double-throw, 60-min cam timer		Automatic Timing & Control

rate is too low, the injector will not operate properly. This problem can be overcome by installation of a glass-lined water tank in the line after the injector (Fig. 8.2). The tank pressure and supply are controlled by an adjustable switch that is set to activate a solenoid at 3.4 atm* (35 psi) and deactivate it at 4.4 atm (50 psi). The glass-lined tank fills rapidly and allows proper functioning of the injector. The tank then becomes a mixing and supply reservoir. If

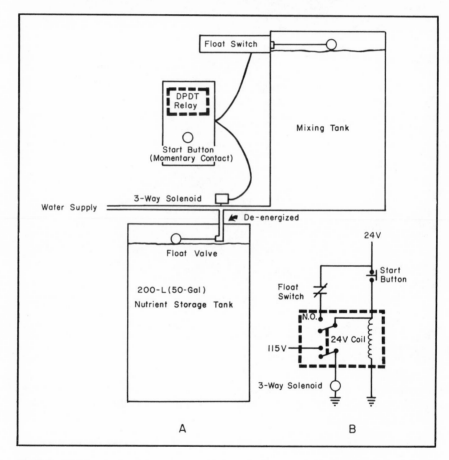

Fig. 8.1. (A) A two-tank nutrient supply system based on gravity flow. (B) Electrical control for a two-tank supply system.

*One atmosphere (atm) = 14.7 pounds per square inch absolute or 0 pounds per square inch gage (psi).

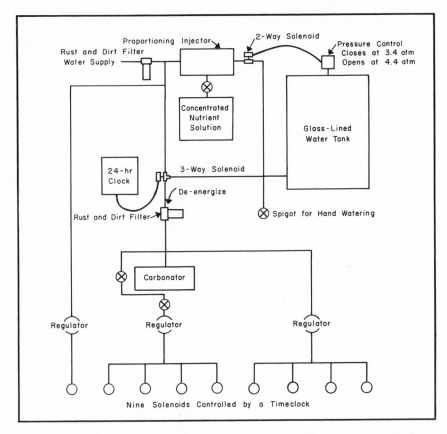

Fig. 8.2. Automatic system for supplying water, nutrient solution, or carbonated water to multiple locations in a controlled environment facility. System may be used for mist and drip irrigation.

desired, plain water can be added by another valve or by a three-way solenoid that bypasses the nutrient supply.

In most environmental experiments, two supplies are recommended—one for nutrient solution and one for water. Leaching with water is usually necessary because of salt buildup in the medium. The salt level can be checked directly with a conductivity meter. By use of separate timeclocks and controls, water and nutrients may be fed into the same distribution system. Use of water provides a simple means for leaching the soil during an experiment. The application of water and nutrient solution 50 percent in excess of need will minimize salt buildup.

Pressurization of the System

For a constant flow to all containers, the watering system must be pressurized. Several methods are available. For the injector system the pressure is provided by the line pressure of the water supply. Since line pressure is usually 6 atm (75.5 psi) or more and may vary during the day or the year, a pressure regulator should be used to maintain a constant pressure.

With the two-reservoir system, researchers have used two methods to obtain head pressure. The tank may be located above the plant level and head pressure is supplied by gravity. A minimum of about 2 m (6.5 ft) difference in height is required. With the other method, head pressure is obtained by pumping the solution. For only a few (20 to 40) tubes, a submersible pump (with a minimum flow of about 800 L per hr at 2 m of head pressure), like those used in water fountains, will work satisfactorily. For large systems a shallow-well water pump should be used in conjunction with a glass-lined storage tank (Fig. 8.3A).

The pump draws the solution from the bottom of the nutrient storage tank and stores it in the glass-lined tank at a pressure of 3.4 atm or greater. A foot valve in the intake line maintains the prime in the pump, and a check valve prevents backflow from the pressurized, glass-lined tank. The pump is controlled by a pressure switch set such that the pump will start at 3.4 atm (35 psi) and shut off at about 4.4 atm (50 psi). If the pressure in the tank drops below 2.4 atm (psi), a low-pressure safety switch energizes a relay and shuts off the pump so it will not burn out. Usually the pressure drops below the minimum only when the nutrient supply runs out. The safety switch can also be connected to an alarm system that signals a malfunction. To restart the pump, one holds open a normally closed momentary switch, de-energizing the relay until the low cutout pressure is exceeded. The schematic of the controls is shown in Figure 8.3B. A pressure regulator beyond the tank is set at 3.4 atm (35 psi). A filter should also be included in the line for removal of rust and other particles.

Distribution of the Solution

The main function of the distribution system is to supply the nutrient solution uniformly over the growing area. Although several systems are available, they fall chiefly into two categories: subirrigation and drip irrigation. In most growth chamber applications, a drip irrigation system is preferred.

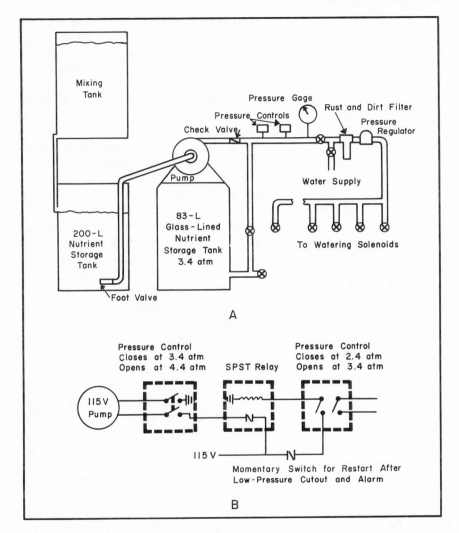

Fig. 8.3. (A) Pressurized supply system. (B) Controls for a shallow-well pump supply system, featuring an automatic cutout and alarm.

Many types of drip irrigation tubes are available for watering plants. They vary in diameter and in design and may be obtained directly from large garden suppliers. To prevent tube clogging, good filters should be placed in the lines. When some of the tubes are not being used, a sliding shutoff valve or a plug (such as a round toothpick) at the end of the tube is useful.

The tubes and manifold should be opaque so algal growth will be prevented. Polyethylene (PE) or rigid polyvinylchloride (PVC) are recommended. Some researchers have found that flexible PVC pipes are not suitable because of the build-up of bacteria on the inside.

The manifold units for the capillary tubes may be purchased, or tubes may be purchased individually, the manifolds made, and the tubes inserted in them. The manifolds can be made by boring holes of .32 cm (1/8 in) diameter in rigid PVC pipe or by puncturing flexible black PE pipe and inserting grommets of .32 cm (1/8 in) internal diameter. The size of the supply line and manifold should be large enough that a minimum uniform pressure of 15 cm of water will be maintained throughout the manifold.

For most small growth chambers, a convenient system includes the necessary number of tubes of the same length and a small submersible pump. When many plants must be watered, a loop system is desirable. In the loop system the supply goes to each corner of the manifold and thereby minimizes pressure differences and evens out the flow. Since the water in the tubing is at a low pressure, the internal diameter must be large. All of our large systems have a high-pressure supply line with a 1.59 to 1.90 cm (5/8 to 3/4 in) ID and a manifold with 2.54 cm (1 in) ID.

After the manifold is built, a check for uniform manifold pressure should follow. A short piece of drip tube is forced into clear plastic tubing (forming a manifold gage) and placed vertically at each corner and in the center of each side of the manifold. Then the watering system is turned on, and the flow valve is adjusted until the head pressure above the drip tube outlets reaches 30 cm as measured by a metric ruler held against the manifold gage. After all the tubes are dripping, the pressure should be measured at all gages. If the pressure on any gage deviates from the average by more than 2 cm, the manifold should be redesigned. If the readings are within limits, all manifold gages except one can be removed. The remaining gage is used for future pressure measurements and adjustments. Flow rates from a sample number of tubes can then be taken. If the manifold pressure is maintained, any number of tubes can then be shut off without affecting the flow rate of the remaining tubes.

On one of the large systems with 3 mm ID drip tubes, we carried out several tests to determine the relationship of time, pressure, and length of the drip tubes to the volume of solution supplied. We found the relationship to be linear, as shown below:

$$Q = 16.2 + 116\,PT/L$$

where Q = volume, ml,
        P = pressure, cm H_2O,
        T = time, min,
        L = tube length, cm.

This equation is satisfactory (R^2 = .95) for a time greater than .5 minute, a pressure from 15 to 45 cm H_2O, and a drip tube length from 30 to 90 cm.

When drip tubes are placed on the soil surface, daily checking of operation is difficult. Therefore, tube outlets should be located just above the soil surface. Placing the drip tubes in a wire holder or inserting them through a drilled hole or notch in a pot label provides a satisfactory means for adjusting the height of the tube outlet. So that flow rates will not be affected during system operation and dripping will not continue after system shutoff, all the outlets should be at the same height, and their height should be the same as or slightly higher than that of the manifold. If only one tube is lower than the others, the tubes and manifold will drain. Then upon system restart the air in the lines must be purged, causing nonuniformity in flow rate. Some tubes may even develop an air lock and not restart. For 15 cm or larger pots, more than one tube per pot will be necessary for uniform wetting of the soil. Special black plastic rings are also available for uniform wetting purposes.

Drip tubes may also supply solution to shallow trays in which potted plants are subirrigated. When subirrigation is used, the trays should have an overflow drain so the depth of the solution in the trays will be limited. Keep the supply line out of the solution to prevent siphoning of the solution through the manifold from one tray to another. In one technique that has worked well, the intake of the overflow drain extends inside the tray to its bottom. The tray then fills with nutrient solution to the level at which the overflow drain passes through the wall of the tray and is maintained at that level by a continuous nutrient supply until it is cut off. When the supply is cut off, the tray drains until it is siphoned empty. With this technique, two things must be kept in mind: the drain opening must be as large as the supply line opening if spilling is to be avoided, and enough solution must be provided to start the siphoning action. One can also subirrigate by maintaining a constant water level with a float valve and setting the pots on sand or some other porous material. If the float-valve method is used, it is critical that the system be level.

In all of the described systems, good filters on water and nutrient supply lines are essential so that clogging of the tubes and nozzles will be minimized. Solenoids should be near the outlets to minimize dripping. If a soil medium with considerable peat is used, it must be given a thorough initial wetting so the seeds planted will germinate quickly and uniformly. Adequate initial wetting is not easily accomplished with either subirrigation or drip irrigation. Mist irrigation immediately after seeds are planted is best.

Automatic Controls

Although the systems described above will operate with hand switches and valves, use of automatic controls will eliminate the need for an operator to be present at a specified time, thereby reducing labor requirements. Controls must be checked periodically so proper operation will be assured.

For control of the daily watering requirements, a standard 24-hour clock, with lugs providing for operation in 15-minute increments and permanent trippers, may be used in series with a 60-minute timer. The 60-minute timer has a minimum on-off time of one minute during its complete 60-minute cycle, and the duration of its "on" time can be increased in one-minute increments. The timer activates one or more solenoids or pumps for supply of the nutrient solution (Fig. 8.4A). Lugs on the 24-hour timeclock are set to allow for one complete cycle (60 minutes) of the timer. The number of 1-minute lugs set on the dial of the timer depends on the desired watering schedule for the day. An example of use of the 24-hour clock with a 60-minute timer follows: The schedule for the day requires watering for 1 minute beginning at each of the following times: 0900, 1200, 1500, and 1800. For that schedule, we would set each of four lugs (15 minutes each) on the 24-hour clock at 0900, 1200, 1500, and 1800 so the clock will operate for 1 hour beginning at each of the four watering times. We would then set one lug (1 minute) on the dial for the 60-minute timer.

A 12-minute timer that provides 12-second increments may also be used in series with a 24-hour timeclock when small volumes of water are required. For example, 12-minute timer is used in misting systems for brief, frequent spraying of the cuttings or the germinating seeds. All controls should be provided with an on-off, timed toggle switch for use when the watering system is checked.

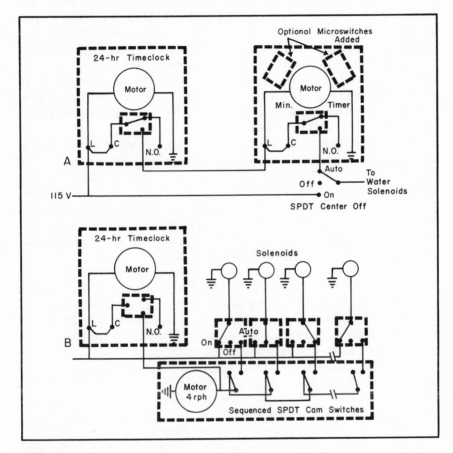

Fig. 8.4. (A) Standard timeclock system suitable for automatically controlling watering at one to three locations. (B) Sequenced cam switches and timeclock for controlling automatic watering at three or more locations for different durations.

In large facilities in which plants in several growth chambers need to be watered or fertilized, the provision of enough solution for uniform flow to all units simultaneously would require too large a supply line. A small line can be used by the installation of several 60-minute timers that are set for watering different locations in sequence. The number of 60-minute timers needed can be minimized by the addition of two more single-pole, single-throw (SPST) microswitches on the dial of one 60-minute timer (Fig. 8.4A). Each switch will activate a separate solenoid. This positioning of switches will automatically provide controls for the watering of three locations at the individual times desired. For more than three locations, a timer with sequenced cam switches can be used (Fig. 8.4B). The cams must be set and then wired in series, as shown in the diagram, so that only one switch will be on at any one time. For more complicated systems, sequence timers or programmable systems can be used. Their use requires some expertise in electronics, and the design of such systems should be worked out for each specific need.

The watering regime of plants can also be controlled by other means. These controls include a photoelectric switch (Koths and Bartok 1975), an electronic or mechanical leaf switch, or a soil tensiometer (Richards and Loomis 1942). A combination of these switches, with or without timeclocks, is also possible. A photoelectric switch is activated by the radiation received by the plants. The evaporation and transpiration of moisture from plants are proportional to the amount of radiation they receive. An electronic or mechanical leaf switch reacts to moisture evaporation and is normally used with a misting system. The electronic leaf switch is activated by the conductivity of the water present, whereas the mechanical leaf switch is activated by a balance of weight of two surfaces. A soil tensiometer measures the moisture tension in the soil. Various sizes are available, but none works satisfactorily in small pots because of the small volume of soil they contain. All of the controls described can be adjusted to compensate for differences in lamps or for differences in frequency of watering. For studies with growth chambers that have a constant level or two levels of environmental controls, the timeclock is the simplest and most widely used control. The special controls described above offer no advantage. A list of materials that have been used in watering systems is included to assist others in the development of their own systems.

Advantages, Disadvantages, Precautions

The use of an automatic watering system offers definite advantages: (1) Water stress is minimized. (2) Labor costs are reduced. (3) A uniform volume is delivered at a predetermined time. (4) Compaction of soil is avoided because of low flow rates. (5) Human error is reduced. (6) Nutrients are readily available when applied in solution. (7) Nutrient solution is conserved because only required volumes are delivered. (8) Splashing is kept to a minimum with the drip irrigation system, thereby minimizing fungal problems.

There are also several disadvantages: (1) Excessive reliance on the system may prevent one from observing plants often enough. (2) Each drip tube must be checked daily during operation, as well as the manifold pressure, the position of timeclocks and switches, and the volume of the supply. (3) Different flow rates for different sizes of pots and plants are difficult to obtain. (4) The positioning of drip tubes at the beginning of each experiment requires special care. (5) Adjustments of flow rates are sometimes difficult. (6) A high initial investment may be involved.

Several considerations are necessary for satisfactory and safe operation of an automatic water system. (1) *Use 3-wire grounded circuits for all components, and comply with all electrical codes.* For safety from electrical shock, use a 24-volt supply under "wet" conditions. (2) Use good solenoids to minimize dripping. For low-pressure systems, use solenoids that are designed for low pressure. (3) Use good timeclocks wired properly to provide safe, dependable, and accurate timing. An on-off-auto, three-position toggle switch can be used for special conditions, but the correctness of its position should be checked daily. (4) Provide large enough supply lines to minimize the pressure drop and subsequent variations in watering among tubes. (5) Include an adjustable hand needle valve, rather than a large gate valve, to adjust the flow into each growth chamber. Maintain a pressure at the tube manifold of close to 30 cm H_2O, using a clear tube to serve as a nanometer. (6) Provide a means of shutting off distribution lines to individual tubes when a full system is not required. The lines can be shut off by a sliding shutoff valve on the end of the tubes or by the insertion of a toothpick or other suitable item in the end of the tubes. (7) Provide an adequate filtering system to keep the lines open and to minimize the clogging of nozzles and tubes. (8) Check the system frequently and check each tube at least daily during a watering cycle. The discharge from extra tubes can be collected and measured for a check of consistent operation. (9) If metal contamination is important, especially in liquid culture, avoid the use of other metal components in the system. Plastic fittings, check valves, and pumps are available. (10) The medium should be lightweight, easily drained, and friable. When using peat mixes do not flood the surface, as flooding compacts it, decreases its volume, and reduces aeration.

A good automatic watering and nutrient system has four basic requirements: a source of clean water and nutrients; sufficient pressure throughout the system; uniform distribution of the solution throughout the growing area; and automatic control of the duration and frequency of application. The keys to the success of any system are flexibility and dependability. Before launching into a large program of automatic watering, one should conduct some preliminary tests to become familiar with the operation of the system.

The primary principle that should be observed in the use of an automatic system is to apply the solution to saturation. The volume of solution required will depend on the medium, environment, and type of plants and containers. When plants are grown in growth chambers, they are usually subjected to higher levels of light, temperature, relative humidity, and carbon dioxide than when they are grown in a greenhouse or field. Plants grown under these conditions grow faster and require more water and nutrients than do their counterparts in a greenhouse or field. With experience the investigator will learn the appropriate volume of water and nutrient solution. Check the volume delivered from a few sample tubes daily. The volume applied should be 50 percent in excess of need to minimize salt buildup.

References Cited

Aljibury, F. K. 1966. Adjustable nozzles simplify irrigation of large container plants. *Calif. Agr.* 20(5):13–14.

Bean, G., and D. A. Wells. 1957. Controlled water applications in glasshouses. *Nat. Inst. Agr. Eng.* 2(2):123–234.

Bean, G.; E. S. Trickett; and D. A. Wells. 1957. Automatic mist control equipment for the rooting of cuttings. *J. Agr. Eng. Res.* 2(1):44–48.

De Vries, M. P. C. 1963. A constant level watering system for pot culture investigations in glasshouses. Australian Commonwealth Sci. and Indus. Res. Agr. Div. of Soils Report. 5 pp.

Geiger, E. C. 1960. Automatic mist propagation. *Flor. Rev.* 126(3258):47–48.

Koths, J. S., and J. W. Bartok, Jr. 1975. Solar timing for pot watering. *Conn. Greenhouse Newsletter* No. 67, pp. 4–7. Univ. of Conn., Storrs.

Larson, R. A., and B. G. Hilliard. 1976. Mat irrigation comparisons. *Grower Talks*, George J. Ball, Inc., pp. 16–23.

Livingston, B. E. 1908. A method for controlling plant moisture. *Plant World* 11:39–40.

Marcussen, K. H., and J. F. Scott. 1965. Application of water in nurseries. *New Zealand J. Agr.* 110:170–179.

Nakasone, H. Y., and F. A. I. Bowers. 1959. Mist box construction for rooting cuttings in Hawaii. Hawaii Ag. Exp. Sta. Cir. 56. 18 pp.

Richards, L. A., and W. E. Loomis. 1942. Limitations of auto-irrigation for controlling soil mixture under growing plants. *Plant Physiol.* 17:223–235.

Robothaln, R. W., and B. T. Pickett. 1972. A programmable nutrient feeding system. Massey Univ. Ext. Paper 20. Palmerston North, New Zealand.

Vanstone, F. H. 1959. Equipment for mist propagation developed at the NIAE. Nat. Inst. of Agr. Eng., *Proc. Assoc. Appl. Biol.* 47(3):627–631.

Wilson, J. D. 1929. A double-walled pot for auto-irrigation of plants. *Bull. Torrey Bot. Club* 56:139–153.

Additional References

Eaton, F. M., and J. E. Bernardin. 1962. Soxhlet-type automatic sand cultures. *Plant Physiol.* 37:357.

Evans, L. T., ed. 1963. *Environmental Control of Plant Growth*. Academic Press, New York.

Furuta, R.; H. P. Orr; W. C. Martin; and F. Perry. 1963. Research results for nurserymen. Auburn Univ. Ag. Exp. Sta. Hort. Series No. 4.

Haahr, V. 1975. Comparison of manual and automatic irrigation of pot experiments. *Plant & Soil* 43:497–502.

Haglund, W. A. 1965. A semiautomatic scale to meter water for greenhouse irrigation. *Phytopathology* 55:473–474.

Hegarty, T. W. 1973. Seedling growth in controlled nutrient conditions. *J. Exp. Bot.* 24(78):130–137.

Kenworthy, A. L. 1972. Trickle irrigation: The concept and guidelines for use. Mich. State Univ. Ag. Exp. Sta. Res. Rpt. 165.

Kramer, P. J. 1975. Uniform watering of potted plants. *Phytotronics Newsletter* No. 10, pp. 22–23. CNRS, Gif-Sur-Yvette.

Krizek, D. T. 1974. Maximizing plant growth in controlled environments. Pages 145–153 in *Proc. XIX International Horticultural Congress*, Warsaw, Poland. September 11–18, 1974.

Mark, A. R., and R. B. Sanderson. 1961. Note on semiautomatic dispenser for water in the greenhouse. *Can. J. Soil Sci.* 41(2):248–249.

Ministry of Agriculture, Fisheries and Food. 1964. Capillary watering of plants in containers. H. M. Leaflet 10, Her Majesty's Stationary Office, London.

New, L. 1975. Automatic drip irrigation for greenhouse tomato production. TVA Bull. Y-94. Proc. of Tenn. Valley Greenhouse Vegetable Workshop.

Post, K., and J. G. Seeley. 1943. Automatic watering of greenhouse crops. Cornell Univ. Ag. Exp. Sta. Bull. 793.

Ridgway, H. W. 1975. Automatic watering with irrometers. *Geiger News* 11(6). Box 283, Harleysville, Penn.

Shumueli, M., and D. Goldberg. 1972. Response of trickle-irrigated pepper in an arid zone to various water regimes. *HortScience* 7(3).

Stanhill, G. 1957. The effect of differences in soil-moisture status on plant growth: A review and analysis of soil moisture regime experiments. *Soil Sci.* 84:205–214.

Stern, J. H.; J. W. White; R. L. Cunningham; and R. H. Cole. Relationship among irrigation media regimes and plant growth. *Plant & Soil* 43:433–441.

USDA. 1970. Growing crops without soil. CA-34-125.

USDA. 1971. Propagation unit for plants. USDA Misc. Pub. 1215, Plan 6101.

USDA. 1976. Building hobby greenhouses. Ag. Info. Bull. 357.

Waxman, S., and J. H. Whitaker. 1960. A light operated interval switch for the operation of a mist system. Conn. Ag. Exp. Sta. Prog. Rpt. 40.

Welch, H. J. 1970. *Mist Propagation and Automatic Watering.* Faber and Faber, London. 162 pp.

Wells, D. A., and R. Soffe. 1962. A bench for the automatic watering of plants grown in pots. *J. Agr. Eng. Res.* 7(1):42.

Wells, D. A., and R. Soffe. 1965. Capillary watering of plants in pots and in beds. Symp. Plant Environ. in Greenhouses, *Acta Hort.* 6.

Went, F. W. 1957. *Experimental Control of Plant Growth.* Chronica Botanica, Waltham, Mass.

Whyte, A. 1960. *Crop Production and Environment.* Faber and Faber, London.

Pests and Diseases

Pests and diseases can cause serious damage to plants grown in a controlled environment. Careful management of the environment is part of an effective control program since certain diseases and pests may be more prevalent under very specific conditions. Once a disease or pest problem has developed, control or eradication becomes very difficult. Therefore a successful preventative program must be an integral part of the general cultural program. The researcher must anticipate the potentially damaging disease and pest problems for the plants being grown. This chapter provides a discussion of control procedures along with a description of optimum conditions for specific diseases and pests that may occur in growth chambers.

Powdery Mildew

Powdery mildew (*Sphaerotheca* and *Erysiphe* spp.) is easily recognized by its characteristic white powdery growth on surfaces of leaves, stems, and sometimes petals (Fig. 9.1). The spores of the fungus are borne on short erect branches that are generally visible with a hand lens (12X to 15X). Mature spores are easily detached from the branches and carried by air currents to surrounding plants where new infections may be initiated. Infection is commonly found on older foliage (Cornell Recommendations for Commercial Floriculture Crops 1974).

The maximum temperature for spore germination is about 22°C (72°F) and germination and infection may occur in relative humidities of 23 to 99 percent (Mastalerz and Langhans 1969). Spore production and germination appear to be correlated with high relative humidity and low temperatures whereas spore maturation is correlated with low relative humidity and high temperatures. Air movement is important after spore maturation in the dispersal of spores. Daytime conditions of high temperatures (27°C, 81°F) and low humidity (40–70% RH), which encourage spore production, germination and infection, alternating with nighttime conditions of low temperature (15°C, 59°F) and high humidity (90–99% RH), which encourage spore maturation, release, and spread, provide the optimum diurnal cycle for powdery mildew. Approximately 3 to 6 days of optimum conditions are required for epidemic development. These conditions encourage powdery mildew epidemics to develop in environmental control chambers if a susceptible plant is being used and the pathogen is present. Such conditions are commonly present in growth chambers, and it is, therefore, advisable to spray plants prior to placement in the chamber with Parnon, Pipron, Benlate, or sulphur. Plants should be closely observed during the experiment for signs of powdery mildew development.

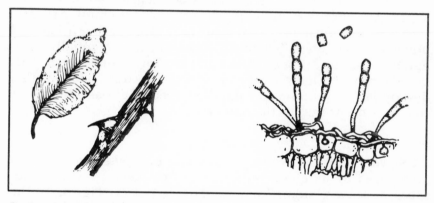

Fig. 9.1. Left: Powdery mildew on rose leaf and stem. Right: Cross section of rose leaf showing chains spores produced on the lesions. (A. W. Dimock collection, Cornell University.)

Rust

The symptoms of rust (*Puccinia, Uromyces,* and *Phragmidium* spp.) are small blisters that appear on the underside of leaves and on stems (Fig. 9.2). The first observable symptoms on the upper leaf surfaces are as chlorotic spots. The epidermis over the blisters or pustules on the lower leaf surface ruptures exposing dark brown or orange dusty spore masses. Pustules rarely appear on upper leaf surfaces. The chlorotic spotting that does appear on the upper surface of leaves is due to death of tissue immediately over pustules on the lower surface.

The fungus spores are transported by air currents or the pathogen may be carried on plants. Free moisture is required before infection can occur and before the spores will germinate. The disease is most severe under cooler temperature conditions (20°C, 68°F). Infection occurs most readily in a temperature range of 10 to 20°C (50 to 68°F) (Langhans 1962). High temperatures above 25°C (77°F) restrict the development of rust by reducing germination of spores, eradicating established infections and killing spores in existing pustules (Langhans 1964, Horst and Nelson 1975). The disease can be prevented by keeping the foliage dry. Should optimum conditions for rust exist in an experiment in which rust susceptible plants are used it would be well to spray the plants with zineb prior to placement in the chamber. Any plants showing rust symptoms after the initiation of the experiment should be removed immediately from the chamber.

Fig. 9.2. Rust on infected leaf with the production of two spore types in lesions. Left: Uredospores. Right: Teliospores. (A. W. Dimock collection, Cornell University.)

Damping-off

Damping-off of seedlings or small plants may be caused by one or more pathogens, usually *Rhizoctonia* or *Pythium,* which may occur separately or simultaneously (Mastalerz 1976). *Fusarium* and *Phytophthora* may also be involved in this complex. Preemergence or postemergence infections may be a part of the disease syndrome that is called "damping-off." Preemergence infection is usually caused by *Pythium* or *Phytophthora* and results in seed decay prior to germination or rot of seedlings prior to emergence. Postemergence infection is commonly caused by *Rhizoctonia* and results in a rot at the soil line after emergence that topples the seedling (Fig. 9.3)

Damping-off pathogens survive in the soil in plant debris saprophytically and as resting bodies, that is, spores and sclerotia. These pathogens are adapted to a wide range of host plants and environmental conditions. *Rhizoctonia* and *Fusarium* are generally favored by warm, moist soils (soil temperatures greater than 20°C [68°F]), whereas *Pythium* is generally favored by cool, wet soil conditions (soil temperatures less than 20°C). These pathogens do not have an air-borne stage, and the spread of these fungi is by mechanical transfer of soil particles infested with mycelia, sclerotia, or resting spores. Therefore, infected plant debris left in controlled environment chambers and dirty flats, tools, trays, and baskets can contaminate new plants. "New" peat used in soil mixes may also carry these fungi (Kim et al. 1975). If the planting medium is steamed or chemically treated and care is exercised to prevent recontamination, damping-off should cause no problem.

Dexon applied as a soil drench gives excellent control of *Pythium*, and Terraclor or benomyl soil drench provides excellent control of *Rhizoctonia* and *Fusarium*. Truban may also be used for controlling *Pythium* but does not appear to be as effective as Dexon. A combined Dexon-Terraclor drench may be used but Terraclor *must not* be applied more than once to a seed flat since it persists in the soil and will result in phytotoxicity. Seed flats with infected seedlings and individual infected seedlings should be removed from the controlled environment chamber to prevent spread of the pathogens.

Botrytis Blight

Botrytis blight (*Botrytis* spp.) is a serious disease of plants and is readily recognized by the brown rotting and blighting of affected tissues followed commonly by the growth of a grey mold fungus on the infected tissues. This fungus attacks a wide variety of plants and is usually identified by the fuzzy grey spore masses that grow over the surfaces of rotted tissues (Fig. 9.4). Sporulation does not develop under dry conditions. Spores are readily dislodged and carried by air currents to other plants.

The spores germinate and produce new infections only when they are in contact with water. This moisture may be from splashed water, condensation, or exudation. Active, healthy tissues are seldom invaded. There may be some exceptions but usually only tender tissues (seedlings or petals), weakened tissues (stubs left after cutting), injured tissues (bases of unrooted cuttings), or old or dead tissues are attacked. This disease is a cool temperature disease. The optimum temperature for development of Botrytis blight is 15°C (59°F). The fungus can survive unfavorable conditions by forming sclerotia in plant debris.

Constant sanitation is extremely critical in reducing Botrytis blight. The removal of dead and dying plants and debris that accumulate in the chambers reduces sporulation of the fungus and thus reduces the spore load in the air. Since Botrytis requires free moisture for spore germination, low humidity reduces the amount of disease development. If high humidities are required for the experiment, good Botrytis control can be obtained by maintaining temperatures above 20°C (68°F). Good cultural conditions for plant growth also aid in Botrytis blight control. Preventative sprays that may be applied to plants prior to placement in controlled environment chambers are captan, zineb, or benomyl.

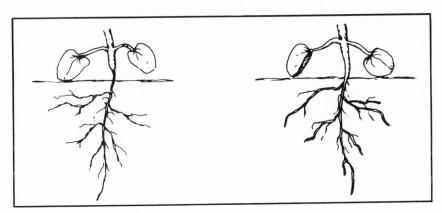

Fig. 9.3. Damping-off symptoms on seedlings. Left: Postemergence damping-off with lesion at soil line caused by Rhizoctonia. Right: Lesions on root tips produced by Pythium. (A. W. Dimock collection, Cornell University.)

Pests

Mites

Mites, which are actually true spiders rather than insects (since they have eight legs), are serious pests on plants. The most widely known mite attacking ornamentals is the two-spotted mite

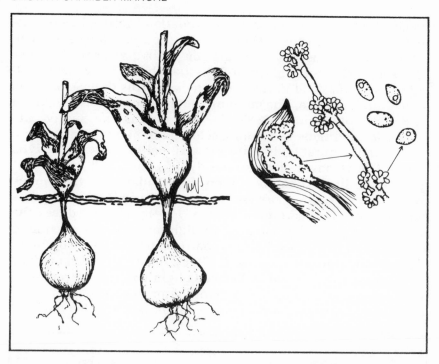

Fig. 9.4. Left: Botrytis blights on tulip leaves. Right: Spores produced on the lesions. (A. W. Dimock collection, Cornell University.)

(*Tetranychus urticae*) (Fig. 9.5). It is sometimes erroneously called the red spider mite. Mites usually feed on the under sides of leaves by means of sucking mouthparts. Pale spots resulting from this feeding may be seen on the upper side of the leaves. The eggs and mites are usually covered by a delicate web that protects them somewhat from contact sprays.

This pest is favored by warm, dry conditions. At nighttime temperatures of 18°C (65°F) and daytime of 25°C (77°F) spider mites pass through one complete life cycle, from egg to adult, in 12–14 days (Mastalerz and Langhans 1969).

Many effective miticides kill spider mites only when they are in the active stages (stages other than the egg). Thus residues toxic for two days or more kill mites hatching from eggs during that period. If the miticide acts only by contact, frequent and regular applications are required to kill when the mites reach the active stage. When repeated treatments are used the risk of phytotoxicity is magnified. In addition, mites may become resistant to

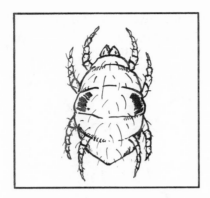

Fig. 9.5. Line drawing of a two-spotted mite

frequently used miticides. Miticides that have exhibited effective control of mites are Pentac, Kelthane, Temik, parathion, tetradifon, Vapona, naleb, and chlorobenzilate. Extreme care must be exercised in using these materials because some are extremely toxic to man as well as to some plants. Vapona for example is toxic to some cultivars of chrysanthemums and chlorobenzilate to some rose cultivars.

Resistance to miticides is sometimes a serious problem. If a particular miticide is not effective, another should be tried until satisfactory control is achieved. There is no advantage to mixing or alternating miticides since resistance to miticides may be achieved more readily. Miticide sprays must be applied to plants prior to placement in the growth chambers because pesticide fumes and spray residues are difficult to remove from the chambers. Soils treated with Temik have been successfully used in growth chambers, but it is advised that treatments be made seven days prior to placement in the chamber.

Aphids

Aphids (*Macrosiphum* and *Myzus* spp. and others) are probably the most familiar and troublesome pest (Fig. 9.6). They have a complex means of reproduction and a series of generations may be produced without fertilization. A portion of the offspring may be wingless while others are winged. Aphids feed on stems, young leaves, and flower buds. They cause distorted leaves and they kill young flower buds or cause distortion of the outer petals. Infested plants are commonly stunted and hardened. Cast skins from the developmental stages of aphids and a sticky honeydew secretion accumulate on plant foliage, and often a black fungus grows on leaf and stem surfaces on this honeydew. Aphids may infest plants in nearly any environment.

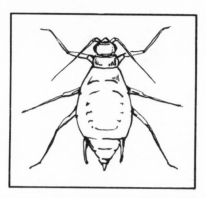

Fig. 9.6. Line drawing of a mature aphid

Chemicals effective in controlling aphids are Malathion, parathion, Vapona, endosulfan, dimethoate, Temik, Zectran, naled, dithio, TEPP, and nicotine. The precautions mentioned previously for using miticides must also be exercised in preventative sprays applied for aphid control.

Whiteflies

Whiteflies (*Trialeurodes vaporariorum*) are very small insects that resemble tiny moths (Fig. 9.7). Adults have wings that are covered with a white dust or waxy powder. They usually feed on leaves of plants and can be particularly damaging to some plants, by sucking sap from the leaves. An objectionable sooty fungus may grow on the honeydew excreted by whiteflies and damage plants by interfering with photosynthesis. Whiteflies may infest plants in nearly any environment.

Parathion or Guthion sprays and parathion, dithio, Vapona, or TEPP aerosols are effective in controlling whiteflies. Temik may also be used. Treatments should all be given outside the controlled environment chambers. Temik treatment should be made at least seven days prior to placement in chambers.

Fungus Gnats

Fungus gnats or manure flies (*Platyura* and *Sciara* spp.) are frequently present in large numbers in wet, highly organic soils

(Fig. 9.8). Larvae of some species feed on fungi and organic matter in the soil, while other species may feed on the roots of plants.

Large populations of adult flies can be controlled by several applications of dithio smoke. Larvae can be controlled with a soil drench of Diazinon or dimethoate. Applications must be made prior to placement of plants in the growth chambers. Diazinon and dimethoate are phytotoxic to some plants. Fungus gnat control is also better achieved by allowing the wet soils to dry out.

Fig. 9.7. Line drawing of mature whitefly

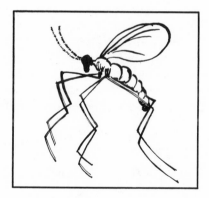

Fig. 9.8. Line drawing of mature fungus gnat

Nematodes

Foliar nematodes (*Aphelenchoides* spp.) may be particularly serious and damaging to plants. These worms emerge from the soil or from previously infested leaves and swim up the stem in a film of water. They enter leaves through stomata and invasion is possible only when plants are wet from sprinkling or syringing. Leaf spots are first discernable on the lower leaf surface as yellowish or brownish areas that eventually turn black resulting in nectrotic spots and defoliation. Although lesions are small at first, favorable temperature and moisture conditions may cause spread until much of the leaf is destroyed. Infection may begin on lower leaves and progress upward.

A simple means of control is to water soil in a manner that avoids wetting the foliage, since a film of water is required for foliar nematodes to move. In addition, cuttings for vegetative propagation should be taken only from the terminal tip of healthy plants. Unlike other nematodes, foliar nematodes do not persist in soil in the absence of living host tissue. Foliar nematodes are effectively controlled by foliar applications of parathion, Vapona, fenthion, and Diazinon or soil applications of demeton or Dasanit.

Algae

Very little has been written about the problems of algae in growth chambers. The blue-green algae require moisture and light for survival. Both of these conditions are generally available in growth chambers and the more water splashed on the chamber floors and walls, the more troublesome the algae. The problems will be exaggerated if nutrient or fertilizer solutions are splashed on floors and walls. Floors on which algae are allowed to grow are treacherous to walk, and serious accidents or falls may occur. Algal growth on the chamber walls reduces reflected light. Algae may also clog the cooling systems.

To reduce algal growth the floors and walls of the chamber should be kept as dry as possible. Window cleaning squeegees should be used to remove water from floors after watering and should be kept available for use in each chamber. Floors and walls should be periodically cleaned to remove algal growth. Shield, manufactured by Puritan Chemical Company, Newark, New Jersey, applied to floors and walls is very effective in inhibiting algal growth for several weeks (McCain and Sciaroni 1965). Biomet 6%, manufactured by M & T Chemical, Inc., Rahway, New Jersey, is also effective. The use of either of these materials for algal control on clay pots may result in phytotoxicity to some plants (Stuart and Cathey 1962). Therefore, these chemicals should be tried on an experimental basis first before using them to control algae in or on pots.

Pest and Disease Control

Prevention

Plant material to be placed in controlled environment chambers should be examined very carefully for undesirable disease or pest problems. Many problems can be avoided by observations made at this time. Should disease symptoms or pests be found, plants must be sprayed, fumigated, or treated prior to placement into the chambers, since many chambers are built to circulate air within the chamber. Fumes from pesticides applied to plants placed in chambers have no way to escape, thus the chamber would act as a fumigation chamber. These fumes may be phytotoxic to plants within the chamber as well as unpleasant for people working in the chamber.

According to the 1972 federal pesticides law, it is illegal to use a pest control product "in a manner inconsistent with its labeling." If a pest is not named for treatment in the labeling of an EPA-registered product, a person violates the law by using the product against the pest. It would not be possible to list all registered usages for pesticides that may be used for pest control on the various plant species that may be grown in controlled environment chambers. If there is doubt concerning the registered usage of a pesticide one should contact a specialist for advice.

Plants should be started from seed directly in the chamber when possible or moved in shortly after emergence. Optimum conditions for pest problems should be recognized (see Table 9.1). Control of diseases and pests within chambers is difficult and involves serious problems. Plants on which diseases or pests are observed should be removed *immediately* so that build-up is not permitted to occur. Plant debris should not be left to accumulate, and hose ends and other watering devices should be kept off of the floor. Floors of chambers must be kept clean and dry to prevent clogged drains and to aid in disease and pest control. Where possible it is advisable to use automatic watering systems that prevent splashing soil. People coming from field and greenhouse experiments should not enter chamber experiments because they may carry pathogen spores and small pests on their clothing and contaminate plant material within a chamber.

All soil used in controlled environment chambers should either be sterilized or pasteurized. Sterilization can be accomplished by use of free steam or by chemical means. Since phytotoxic materials often accumulate in freshly steamed soil and since 100°C

(212°F) is not required to kill all harmful pests and pathogens found in soil, there is increased interest in pasteurizing soil at temperatures of 68°–71°C (155°–160°F). Harmful pests are killed at 60°–62°C (140°–145°F). Pasteurized soil can be used immediately after cooling.

Eradication

Affected plants must be removed from the chamber as soon as detected to prevent disease or pest build-up, which is often explosive. Optimum conditions for a specific pest frequently exist in the chamber and the pest thrives.

Sprays and fumigants must not be used inside chambers, because pesticide fumes are difficult to remove from the circulated air of chambers. These fumes can be extremely toxic to personnel working in the chambers as well as to the plant material in the chamber. Plants can be removed from chambers, sprayed or fumigated, and then returned to the chambers after treatment.

It is very difficult to decontaminate chambers once diseases or pests have been allowed to build up in large populations and become established in the circulation system of the chamber. The best course is to shut down the chamber for 4 to 6 weeks so that no host plant material is made available for continued survival of the pest in the chamber. Walls, carts, and chamber floors should also be scrubbed with a good disinfectant such as Amphyl or Chlorox solutions.

Effect of Pesticide on Plants

Specific pesticides may influence plant growth in ways other than the more commonly detected phytotoxic reactions associated with pesticide treatments. Specific cultivars of some crops may be particularly sensitive to certain pesticides. Some cultivars may be more sensitive under specific environmental conditions. For example, chrysanthemum cultivar 'Shasta' is extremely sensitive to Vapona, especially if the treatment is made under high temperatures.

There is increased interest in the use of systemic pesticides such as Benlate and Temik. Because these chemicals are translocated within the plant, they influence plant metabolism. Benlate is known to increase plant growth in some ornamental plants, such as azalea, under certain conditions (Horst, unpublished). Temik is also known to influence the production of phytoalexins (Steiner 1976). (Phytoalexins are chemicals produced by some plants in response to invasion by a pathogen; they are antagonistic to the pathogen.) The investigator must recognize and evaluate these external influences on experiments run under controlled conditions.

Table 9.1. Optimum conditions for pests and diseases

Pest	Optimum temperature	Optimum moisture
Powdery mildew	15°C (59°F)	99% RH
Rust	20°C (68°F)	free moisture
Damping-off		
Pythium	below 20°C	moist soils
Rhyzoctonia	above 20°C	
Botrytis blight	15°C (59°F)	free moisture
Mites	25°C (77°F)	dry
Aphids	wide range	wide range
Whiteflies	wide range	wide range
Fungus gnats	wide range	moist soils
Nematodes	wide range	free moisture
Algae (require light)	wide range	free moisture

References

Cornell Recommendations for Commercial Floriculture Crops. Part II. Disease, pest and weed control. 1974. New York State Coll. Agric. and Life Sci., Cornell Univ., Ithaca, N.Y. 56 pp.

Horst, R. K., and P. E. Nelson. 1975. Diseases of chrysanthemum. Cornell Univ. Ag. Info. Bull. 85.

Kim, S. H.; L. B. Forer; and J. L. Longenecker. 1975. Recovery of plant pathogens from commercial peat-products. *Proc. Amer. Phytopath. Soc.* 2:124 (abstr.).

Langhans, R. W., ed. 1961. Carnations: A manual of the culture, insects and diseases, and economics of carnations. New York State Flower Growers Assoc., Inc. 107 pp.

Langhans, R. W., ed. 1962. Snapdragons: A manual of the culture, insects, and diseases, and economics of snapdragons. New York State Flower Growers Assoc., Inc. 93 pp.

Langhans, R. W., ed. 1964. Chrysanthemums: A manual of the culture, diseases, and insects, and economics of chrysanthemums. New York State Flower Growers Assoc., Inc. 185 pp.

Mastalerz, J. W., ed. 1971. Geraniums: A manual on the culture, diseases, insects, economics, taxonomy, and breeding of geraniums. Pennsylvania Flower Growers. 350 pp.

Mastalerz, J. W., ed. 1976. Bedding plants: A manual on the culture of bedding plants as a greenhouse crop. Pennsylvania Flower Growers. 514 pp.

Mastalerz, J. W., and R. W. Langhans, eds. 1969. Roses: A manual on the culture, management, diseases, insects, economics, and breeding of greenhouse roses. Pennsylvania Flower Growers, New York State Flower Growers Assoc., Inc., and Roses, Inc. 331 pp.

McCain, A. H., and R. H. Sciaroni. 1965. Algae control in the greenhouse. Flor. Rev. 137(3545):28.

Steiner, P. W. 1976. Alfalfa phytoalexin degradation by *Stemphylium botryosum* in relation to pathogenicity and virulence. Ph.D. thesis, Cornell University, 99 pp.

Stuart, N. W., and H. M. Cathey. 1962. Control of algae on pots with "36-20." Flor. Rev. 120(3369):17.

Chapter **10** WILLIAM A. BAILEY, DONALD T. KRIZEK,
AND HERSCHEL H. KLUETER

Preparing Specifications

There are many ways to write specifications. Each manufacturer, university, or private business has a unique method of writing them. A specifications writer should refer to the guidelines for controlled environment enclosures published in 1971 by the AIBS Bioinstrumentation Advisory Council entitled "Controlled Environment Enclosure Guidelines." Its workbook approach to writing these specifications enables the manufacturer and the purchaser to understand what is ordered. Specifications are legally binding and, if properly written, will help to avoid contractual problems after the units are put into operation. By supplying the manufacturer with all of the relevant information, the purchaser stands a better chance of obtaining a growth chamber that satisfies his requirements.

There are three types of specifications—performance specifications, engineering specifications, and a combination of both. Performance specifications describe the performance expected from the unit, whereas engineering specifications describe in detail the parts to be used and the methods of assembly. Use only one type of specification when preparing a bid invitation or a purchase contract. If two types are listed, the contractor has the privilege of selecting the type that best suits his purposes. In general, the performance specification that includes a guaranteed performance of one year or more will obtain the best operating plant growth chamber at the lowest cost. There are several references available to aid the writer of growth chamber specifications (*ASHRAE Handbook of Applications* 1974; *ASHRAE Handbook of Fundamentals* 1977; Downs 1975; Morse and Evans 1962; Pescod et al. 1962; Chouard and de Bilderling 1972).

The first items to be considered are the total cost of each chamber or group of chambers, the external dimensions of the chamber, the area and height of the plant bed, the number of chambers needed for the particular experiment or plant production, and the desired location of these chambers. One or more of the manufactured growth chamber models may satisfy these needs without unique specifications.

The growth chamber with the lowest cost per unit area of growing space will probably be a standard model from the assembly line. Standard sizes can also be replaced easily when they become obsolete or need major repairs. Always build flexibility into the chambers, where possible, because every chamber will become obsolete in time and will need to be changed. If a small part of a chamber can be changed, it will be less costly than putting in a whole new chamber. However, if you need a special size or type of chamber, write your specifications accordingly.

The expected use of the chamber should be briefly described in the specifications. Also name the crop to be grown and indicate the minimum growing area and height that you will need in your experiment. If corn or other tall plants are to be grown, specify extra height in the chambers. A general description of the planned location of the chamber should be supplied. This description should include information on the availability of air conditioning in the building, the type of electrical service (voltage, hertz, amperage, number of phases, and type of hookup), the type of water available (hot, cold, distilled, deionized, or steam), availability of compressed air or vacuum systems, and location of drains. Other information about the area that should be given includes the ceiling height, the type of floors, and the width of doors, windows, and hallways through which the unit must be moved. If the chamber is to be located on an upper floor or in the basement, be sure that the doors and the elevators, if any, are large enough to take the largest section of the equipment. It is quite costly to hire a crane to lift a unit up the outside of a building and then bring it through a window or a hole knocked in the wall. If the refrigeration equipment, storage tanks, nutrient tanks, or other equipment are installed on a different floor or in a different room from the growth chambers, a schematic or blueprint of the building should be furnished and the ambient temperature and humidity conditions where the chamber will be operating should be reported. This general service information is not a binding part of the contract, but it will help both the user and the manufacturer to avoid problems during the assembly, delivery, and operation of the chamber.

Choosing a manufacturer. The established manufacturer will probably be the best source for obtaining a plant growth chamber. However, a new manufacturer may have a better idea, may be able to cut some manufacturing costs, or may be able to service chambers locally at a lower cost than could the established manufacturer. Therefore, a new manufacturer should have a right to bid on any new chamber. Shop around for a growth chamber just as you would for any large appliance. However, keep in mind that a plant growth chamber contains specialized refrigeration and lighting equipment and, therefore, should be manufactured and serviced by experts.

Economics of purchase. The lowest-cost growth chambers are the standard ones that are manufactured on an assembly line. By building 50, 100, or 1000 such units at one time, the manufacturers can reduce their cost and yours to a much lower level than if they have to make a chamber to order. A few custom manufacturers who make chambers to order could be approached to bid on any special item.

Overspecification. The problem of overspecification for growth chambers has been a factor in creating high costs for users. When users do not know the regimes with which they will be working, they often specify a greater range and tighter control than is necessary. Since such precision raises the costs unnecessarily, as a general rule, realistic specifications should be prepared.

Technical Requirements

Construction

The wall materials and finish should be specified. White, baked-on enamel, epoxy paint, aluminized Mylar, or other reflecting surface normally is used on the inside wall and a natural finish or an enamel on the outside walls. If the laboratory has a particular interior decoration scheme and a certain color is needed, specify that color. The walls are normally made of foam or fiber glass insulation, with wood or metal framing. Foamed-in-place polyurethane has greater insulating value per inch of thickness than does fiber glass or other insulation materials.

Dimensions. When specifying the size of the growth chambers, allow adequate space around each chamber for moving it into place or taking it out for repair work. Leave space behind each unit so that service personnel can get to it easily. This extra space will also leave a place to run plumbing and electric lines and to install working lights and electrical service outlets for handtools. Specify the minimum inside dimensions, which should include the length and width of the chamber floor and the height of the growing area. Also specify the length and width of the floor and the clear height of the chamber. These dimensions should be supplied for reach-in and walk-in growth chambers.

Plant bed. Some plant beds are stationary and others are movable. Although most plant beds are left in place, it is convenient to have some way to vary the height to compensate for a decrease in light intensity or to maintain the same distance from the top of the plant to the lamp bank. Specify whether you wish to adjust the plant bed with a hand crank or by other means, e.g., shelf brackets. Some laboratories use carts both for growing the plants and for moving them between the greenhouse, growth chamber, and laboratory.

Doors and seals. The seal around the door prevents the leakage of light and carbon dioxide or other gas into or out of the chamber and also contains the conditioned air. The magnetic seal that is used on household refrigerators appears to work best. However, specify that the seal be opaque to light. If an inspection window is desired in the growth-chamber door, so specify; but remember that the cover over the window can be accidentally opened, allowing light to reach the plants during the dark period.

Access ports. It is usually desirable to have one or more access ports in each chamber to allow instrument leads to be placed inside the chamber and to provide a means for adding water, CO_2, or other gases to the growing area. Care should be taken to make these ports light-tight to avoid photoperiod problems.

Barrier. Some chambers employ a barrier and others do not. There are good reasons for each arrangement. The use of a barrier between the lights and the growing area reduces the heat load from the lamps and the light levels. In a chamber without a barrier, the airflow normally goes from below the plants up through the hot lamps and back to the cooling coils. This airflow pattern insures that the hottest air reaches the cooling coils and the heat is carried away from the chamber.

In a chamber with a barrier, two areas need cooling—the light cap and the plant growing area. In some chambers the refrigeration unit takes care of the lamp bank and the growing area. In other chambers, the refrigeration unit only takes care of the growing area. The lamp bank is then ventilated to the outdoors or cooled by an auxiliary cooling system. If cold temperatures are to be used in the growing area, it will be necessary to have a barrier between the lights and the growing area or to use special outdoor lamps. Fluorescent lamps operate at maximum efficiency at a wall temperature of 38°C. Therefore, any temperature below this level reduces the light output. When the barrier is clean, it reduces the light transmission about 10 percent. If it becomes dirty, as it often does, the reduction will be greater. Barriers should be easy to remove for cleaning and for replacement of lamps.

Floors and drains.   A suitable floor may have been built into the floor of the building originally. If not, a new floor should be made of galvanized metal, stainless steel, or other durable substance. If caustic chemicals will be used in the chamber, it may be desirable to specify stainless steel. For safety, walkways should be coated with an abrasive.

The drain for the growth chamber is normally placed in the floor. The specification for the location of the drain should be such that it will be easy to reach and clean out without removing the plants.

In a chamber where the soil will be watered by spray nozzles or individual watering tubes, it is desirable to have some way of draining away the excess water from the container. The drain should be at least as big as that for a kitchen sink to prevent stoppage from soil or plant debris. It should also have a water-sealed trap to restrict gas flow into or out of the chamber. An external floor drain is needed under the chamber in addition to the one inside. The floor around and under the chamber should be sloped to this external floor drain.

Refrigeration Systems

Three main types of refrigeration systems are used in growth chambers today: direct expansion with a normal, or on-off, control system; direct expansion with a modulated control system; and secondary cooling. The first system is probably the most economical to purchase initially. The second and third types are the most desirable for a growth-chamber operation because they have small temperature and humidity fluctuations. The temperature approaches a modulated line and stays at the preset temperature.

For a small installation, the direct-expansion modulated system normally is the most desirable, but if a whole building is to be full of growth chambers, a secondary cooling system may be desirable. In this system one or two large compressors, which are more efficient than small compressors, are used to cool an anti-freeze solution that is circulated through the building, and each chamber is connected to this central supply of cold solution.

There are several different direct-expansion, modulated refrigeration systems. Some of them use solenoids to control the flow of the freon refrigerant, others use modulated valves that open or close as needed to allow the correct flow rate of refrigerant. Valves should be easily accessible for servicing.

Capacity. The compressor capacity, or size, will be determined by the plant and radiation heat load, the wall and floor area, the volume, and the inside and ambient temperature. Here the specifications should be directed to performance so that the manufacturer must furnish the correct size of compressor. The compressor that is to operate in a hot area must be larger than one that is to operate in a cool area to provide the same inside temperature. Therefore, the location of the compressor is quite important and should be included in the specifications. If the compressor is to be placed on the roof or outdoors, so indicate, because most compressors are not designed to operate at temperatures below 13°C. If such operation is necessary, a special year-round compressor or condensing system must be specified.

The size of the cooling coil and the capacity of the blower are important. Most refrigeration systems are designed to have an 8°C temperature difference (TD) between the air entering the coil and the air leaving the coil. This temperature difference helps remove moisture; if a high humidity is needed in the chambers, the coil must be larger than the one normally used in order to reduce the TD. Performance specifications should be written so that the manufacturer must furnish the correct coil and blower to meet the chamber's requirements. If you try to specify the coil size and the TD, the coil may not operate to meet your performance requirements. The material of the coil should be specified, however, because if you are going to use pollutants, the coil should be either made of stainless steel or epoxy-coated metal. Pollutants such as sulfur dioxide corrode and destroy aluminum coils.

If the lamp bank is to be cooled separately from the growth chamber's refrigeration unit, the coil sizes and the refrigeration capacity must be matched to the heat load created by the lamps.

Standby refrigeration. If it is necessary to keep a plant growth chamber operating continuously, a standby refrigeration system should be purchased initially. The capacity of the standby system must be as large as that of the original unit, and the piping must be installed in parallel so that the system can be put into operation automatically by solenoid valves, or manually by opening the closing the plumbing valves. If an installed standby system is too expensive, it would be desirable to have a new compressor and other refrigeration parts for a standby system in stock so that they can be installed immediately. Provision for a standby system is important if several identical chambers are to be used at one laboratory or group of laboratories. Standby electric power generation equipment should be installed. It should be sized to operate the refrigerator equipment and the low level lighting for photoperiod control along with electric controls, recorders, and other necessary auxillary equipment (ASAE 1977, Campbell 1965).

Lighting.   There are many types of lamps available for lighting plant growth chambers (Canham 1966, Carpenter and Moulsley 1960). If you will need a special type of lamp or control—for example, in order to turn the lamps on for a few seconds in the middle of the night—this requirement should be included in your specifications. Lamp banks can be specified either by performance or by the number and types of lamps. If you want a certain irradiation level—for example, quantum flux density, or lumen output—so specify. If you specify an irradiation level, it should be that required after about 2000 hours of operation. The reason for this time specification is that most lamps have a maintenance or darkening factor that is related to the age of the lamp.

It may be desirable to have a lamp bank canopy that can be completely removed so that new lamps can be installed at a later date if a better type becomes available. However, no lamps should be installed that would produce more heat than those in the original installation, because the capacity of the chamber's refrigeration system will be designed to handle the heat load created by the lamps in the original installation. Of course it is entirely possible that more efficient lamps than we now have will be developed to provide more irradiation in the visible and plant-response range of 400 to 750 nm, with less heat.

Normally the desired ratio of red lighting to far-red lighting is obtained by using incandescent lighting to supplement high intensity lighting. Where possible photosynthesis lighting should have controls separate from the photoperiod controls so the incandescent lamps and the fluorescent lamps can be operated simultaneously or independently.

Most manufacturers give lighting specifications in terms of number, loading, and type of lamps. Intensities specified are generally those obtained for new lamps near the lamp bank and are therefore higher than those recorded at practical growing distances or after the lamps have been operated for several months.

Most plants can be grown satisfactorily under 32–40 nE cm^{-2} s^{-1} of photosynthetically active radiation (PAR). This is equivalent to approximately 2000–2500 ft-c of cool white fluorescent light. For most plants supplemental incandescent lighting should be added in the ratio of at least 25 watts of incandescent light to 100 watts of fluorescent light.

A high-temperature cutout should be provided in the lamp chambers so that the temperature will not rise above a certain preset level if the refrigeration or the cooling fan goes off. If the lights are not cut off automatically, the plants may be killed or the insulation on the electric wires may be melted and a fire may start. At least 10 ft-c of photoperiod lighting should be maintained when the growing lamps are cut off in order to avoid problems caused by undesirable phytochrome or daylength control (Hendricks and Borthwick 1963).

Carbon Dioxide Control Systems

In preparing CO_2 specifications for a growth chamber, state whether the chamber will be sealed to retain CO_2 and to exclude the outside air. To fulfill this requirement there should be no fans to add outside air, or if these fans are required for experiments, a means should be provided for sealing the air inlets and outlets. The CO_2 should be fed into the growth chamber in such a way that it will be diffused in the area of the circulation fan before being introduced to the plants. This diffusion will eliminate any clouds of high-density CO_2. The control system can be used to add or remove $C\bar{O}_2$ during either the light or dark periods, or during both (Bailey et al. 1970).

Ventilation System

More controversy occurs between growth chamber manufacturers and users about the direction of airflow than about almost any other factor of the system. Air must be moved to carry heat away from the plant, to carry carbon dioxide to the plant, and to maintain minimum temperature gradients, both vertical and horizontal, within the chamber. Any airflow rate and direction that can accomplish these objectives will probably be satisfactory.

Air movement. The air moving over the plants interacts with light and heat radiation, relative humidity, and plant transpiration to form microenvironments throughout the plant growing area. Therefore it is necessary to hold the airflow constant over the whole growing area. To do this the rate of airflow must be high enough to pressurize the air and to distribute it evenly across the plant bed. If the rate of airflow is too low, there will be too great a temperature rise and the air will move in puffs, or along channels. If it is too high, it may dry out or physically damage the leaves. The airflow through the plants should be at least 15 cm sec^{-1} (30 fpm) and not over 50 cm sec^{-1} (100 fpm). Avoid specifying air movement in air changes per minute since one air change per minute in a one-foot-cubed chamber will mean .5 cm sec^{-1} (1 fpm) velocity, while the same air change in a ten-feet-cubed chamber will mean 5 cm sec^{-1} (10 fpm) velocity.

Makeup air. Unless provisions for carbon dioxide are available it will be essential to specify a certain amount of outside makeup air. A small positive blower can be used to force this air into the chamber. A small amount of makeup air will bring in carbon dioxide, and exhaust air will help carry out impurities such as ethylene and ozone. Remember, however, that the cost of conditioning every cubic foot of makeup air will be high, because the air must be brought to the same temperature and relative humidity as the air inside the chamber. Therefore keep the percentage of makeup air to a minimum. Of course, some makeup air is required for normal plant growth in chambers that rely on ambient carbon dioxide.

Filtering. If the growth chambers are to be used to study air pollution, it will be necessary to specify that filters be provided to remove all air pollutants from the makeup air. Most of the pollutants can be filtered out by using first a coarse furnace filter to remove the particles of dirt and debris, and then a charcoal or potassium permanganate filter to remove objectionable gases. Fine or absolute filters are available to eliminate bacteria when such a requirement exists. Remember that extra power is required to force the air through the small pores of such filters. To prevent the infiltration of polluted ambient air into the chamber, the chamber must be operated at a positive pressure.

Utilities

Electricity. The electrical supply should be specified in volts, hertz (cycles), number of phases, and types of hookup. (Example: 208 V/60 Hz/3 Ph/4-wire "Y".) If an electric supply is already installed, the available amperage should also be indicated.

Water, steam, pressurized air, and vacuum. The size of water supply pipes and of drains and the distance of each from the intended installation area should be specified. If distilled or deionized water is available in the area where the plant growth chamber will be installed, the available flow in gallons per hour or liters per minute should be specified. Distilled or deionized water can be used for humidification, for nutrient solutions, for leaching of soil, and, if available in sufficient quantity, for the cooling tower.

If steam is available, indicate the size of line and its pressure. The manufacturer will need this information if steam is to be used for heating or humidifying the chambers.

If pressurized air and vacuum systems are available, describe them so the manufacturer can make use of these systems if necessary. Pressurized air used in controls and for plant experiments must be free of oil and other impurities.

Service Requirements

Installation

Specify who is to do the installation—the manufacturer or the purchaser, or both—and how much each is expected to do. If the purchaser has an installation crew available, they might be able to install the equipment more economically than the manufacturer could.

Standards of Performance and Acceptance

Although a requirement for the provision of exact tolerances could increase the cost of the original bid, it may be desirable to list the tolerances that you will accept in your chamber. Specify who will do the testing, the type of test equipment, and the acceptable variations. No two chambers operate exactly alike, so the manufacturer's specified data may not describe the way that your own unit will operate.

Guarantee

The guarantee on a product is only as good as the manufacturer and only as binding as the accepted specifications. Most guarantees become void if the product is misused, so care should be taken to handle the growth chamber and its equipment properly. Most warranty periods are for one year from date of acceptance of the product.

Delivery Requirements

Most growth chambers are shipped by commercial carriers and are delivered by the truck driver to the receiving area. If specific requirements must be met—delivery to a certain building on a certain date, or knocking the chamber down to a certain size or weight so that it can be accommodated by the elevators or doors through which it must be moved—such requirements should be in the original specifications. If the costs of a special procedure will be excessive, the manufacturer can take an exception and list these costs separately. Then the purchasers will be able to determine how to handle the cost.

Modifications

Minor modifications can usually be made by either the manufacturer or the purchaser. When such modifications are necessary, it is often desirable for the purchaser to modify a standard unit at the place of use. It is usually faster and more economical to obtain standard growth chambers on approved purchase schedules than it is to obtain modified chambers that must go out on bid.

Sample Specifications

The following sample specifications are published for your guidance only. They are not intended to be used to purchase any particular chamber for any particular use, but rather as an aid in writing specifications that will fit your particular needs and that will give all of the necessary information.

Specifications for Controlled Environment Chamber Service Requirements

The research conducted in the ____ laboratory at ____ requires accurate and reproducible controlled environments for physiological studies on horticultural crops such as lettuce, tomatoes, petunias, marigolds, birch, crabapples, and selected experimental species, and on field crops such as clover, alfalfa, soybeans, wheat, oats, and corn.

For the proposed physiological research, accurate control of light intensity, day length, temperature, relative humidity, carbon dioxide (below and above atmospheric concentration), and airflow are needed. In addition, the environment must be scrubbed for excessive concentrations (above atmospheric) of undesired gases such as ethylene, ozone, sulfur dioxide, and carbon monoxide.

For purposes of definition, the words "growth chamber" will mean the total system, and the words "growing area" will mean the volume within the growth chamber where plants are grown. "Growth height" is the maximum distance between the light barrier or lamp bank and the floor on which the plants sit.

A minimum of four separate but identical plant growth chambers are required. To satisfy the research requirement of the laboratory, each of the controlled-environment chambers must meet the following performance and technical requirements.

I. Physical Construction

A. Materials

Since the growth chamber is subjected to high humidities, it must be constructed of rustproof materials such as aluminum, stainless or coated steel, or plastic.

B. Size

To provide the needed capacity the interior dimensions of the growing area for the plants must be at least ___m wide by ___m front to back, and the growth height must be at least ___m. So that the growth chamber can be moved through existing doors, its exterior dimensions shall not exceed ___m front to back by ___m high.

C. Interior and Exterior Finish

To provide adequate light reflectance the interior walls of the growth chamber must be either highly reflective material or painted white with a baked-on enamel or epoxy paint, or a coating of equal quality. The surface shall be of material or coating that will not corrode, fade, yellow, or otherwise lose its reflective ability after several years of service. The finish on the outside walls shall be ___.

D. Insulation

The insulation shall be of closed cell foam polyurethene and shall be covered with metal or fiberglass skins. To prevent cold spots or hot spots there must be no metal or other material having high thermal conductance extending continuously from the interior walls to the external walls. The walls shall not conduct more than 0.10 Btu $(0.57 \text{ Wm}^{-2} \cdot \text{K})$ per hour per square foot per degree Fahrenheit of temperature differential between the inside and the outside of the growth chamber. (To convert from English units to SI units multiply by 5.674466.)

E. Floors

The growth chamber must have a floor (or the growth chamber floor may be built into the purchaser's building floor—if so list the dimensions, structural characteristics, and location of drains in the specifications) that will provide for the collection and removal, by means of a drain, of condensed water or water spilled from the watering of the plants and of dirt, leaves, and other small particles of debris that are products of the experimental process. Since the floor will be exposed to water, soil, and plant matter, it must be made of material that will not rust, corrode, or otherwise deteriorate under normal use.

The bench (floor) of the growing area must support a load of at least 244 kg/m² (50 lbs/ft²) evenly distributed across the bench. Since it is desirable to adjust the location of the various sizes of plants in relation to the light source, the bench height must be adjustable from the floor to within 30 cm of the light barrier. In addition, since it is desired to adjust the plant location vertically without altering the environmental conditions, the bench must be adjustable without disturbing the plants.

F. Doors and Seals

To allow adequate access, at least one hinged access door must be provided in one side of the growth chamber. The door opening must be at least 60 cm wide and 90 cm high. The door seal must be light-tight. To allow for observation of the plants in the growing area without exposing them to the ambient atmosphere, the access door must have a window that can be made light-tight by adding a hinged cover.

G. Lamp Bank

The light intensity will be kept at a constant level in the growing area. Since the light output of fluorescent lamps varies with temperature, it is necessary to keep the lamp wall at a constant predetermined temperature (38°C ± 5°). Therefore, the lamp bank must be isolated from the growing area, and a separate system must be provided for cooling the lamp bank. One specification might read as follows: the lamp bank shall contain 24 1500 maT-10, 1.2 m cool white fluorescent lamps and 12 100 watt, 130 volt incandescent lamps. The lamp bank shall be completely sealed (no transfer of gas with a pressure differential of plus or minus 1.25 cm [one-half inch] of water) from the growing area by means of a transparent barrier placed between the two.

The airflow in the light chamber shall be of such quantity distributed in such a manner that the temperature of each fluorescent lamp will not vary more than 5°C from the average temperature of all the lamps. The lamp temperature shall be measured at the bottom and the midpoint of the tube.

(An alternate specification for the lamp bank could read: It is desirable to keep the light intensity at a constant level in the growing area. Because the light output of fluorescent lamps varies with temperature, it is necessary to keep the lamps at a constant temperature of 38°C ± 5°; therefore, shields shall be placed around the lamps and the air shall move up or down past the lamps in such a way that the lamps are held at a constant temperature for maximum light output.)

H. Other Access Areas

A minimum of two access holes must be supplied to the growth chamber for water, instrument leads, and gas supplies. Each hole must be at least 5 cm in diameter, or equivalent area. These holes must be capable of being completely sealed against gas and light transfer, as described above.

I. Growth-Chamber Sealing

The mixture of gases in the growth chamber will be altered (mainly the proportion of carbon dioxide) from that normally found in the atmosphere. The growth chamber must be sealed to restrict the transfer of gas either in or out of the chamber. The seal must be capable of maintaining a differential pressure of plus or minus 1.25 cm of water between the growth chamber and its surrounding air. For controlled carbon dioxide studies, the air system must be capable of being sealed to lose no more than 10 ppm of CO_2 per minute at a differential of 500 ppm of CO_2.

II. Environmental Control

Sufficient control of temperature, humidity, light, and carbon dioxide must be maintained to insure the environment required for the research to be conducted in the environmental growth chambers.

A. Temperature

The temperature control system must maintain the exhaust air in the growth chamber at any set point within the range of 5°–45°C, with a maximum differential of 0.5°C from the set point. It must do so under all possible combinations of levels of the other environmental conditions, when the lights are on and when they are off, even though the temperature of the ambient air surrounding the growth chamber may vary from 0°–50°C.

The variation from the top to the bottom of the growing area shall be no more than 1°C, and the variation across the growing area on a horizontal plane shall be no more than 1°C.

The exhausting of heat into the room that contains the growth chamber will be undesirable. Therefore, the refrigeration system must be provided with a water-cooled (or air-cooled) condenser outside of the room. The contractor shall install this remote unit. Special insulation and heaters shall be added to allow winter operation of either the water-cooled or air-cooled refrigeration systems.

B. Humidity

The humidity of the exhaust air in the growth chamber shall be controlled to within 5 percent of the set point between 50 and 90 percent relative humidity. The controller must be synchronized with the day-night timeclock for operating the chamber. To test the humidity controls, the bench will be filled with round No. 2 open-topped cans that are half full of water. The test conditions will be maintained long enough for the water temperature to stabilize. The cans will be staggered (in a triangular pattern) and touching each other to insure that about 90 percent of the growing bed surface is covered by water. The evaporation from these cans of water will be approximately equal to the transpiration from the plants when the bench is fully covered.

C. Light

After 2000 hours of operation, the fluorescent lighting system must provide at least 40 nE cm^{-2} s^{-1} of photosynthetically active radiation (PAR) at a distance of 60 cm below the transparent ceiling barrier or the open lamps. The intensity of the light at any point at this level, in a horizontal plane extending to within 15 cm of the walls, must not vary from the average intensity at this level by more than 10 percent. Three on-off contactors must be provided for controlling the incandescent lamps and one-third of the fluorescent lights. The contactor must be controlled by a 24-hour timeclock. (If fluorescent and incandescent lamps are to be used, state the number, wattage, voltage, and ratio of fluorescent watts to incandescent watts; do the same if mixed, high-intensity discharge lamps are to be specified.)

D. Airflow

The growth chamber must have a system for recirculating the air within the growing area. The airflow can be from top to bottom, bottom to top, or side to side. The velocity of the air in the chamber must not exceed 50 cm/sec^{-1} or be less than 15 cm/sec^{-1} through the plant growing area. The velocity of airflow, as measured in a horizontal plane 60 cm below the transparent ceiling barrier to within 15 cm of the walls, must not vary more than 5 percent from the average velocity in that plane when the chamber is empty or when it is fully loaded with evenly spaced plants. The airflow system shall be equipped with an inlet to add fresh air, at any predetermined rate from 0 to 300 liters/min, for CO_2 makeup.

E. Control of Carbon Dioxide and Other Gases

The CO_2 sensor will take a sample of air from the chamber as it leaves the growing area and shall control it at any preset level, ±20 ppm of CO_2, over a range of 200 to 2000 ppm. The controller shall be of the infrared-gas-analyzer type or a unit with equal or better performance. The CO_2 controller can be purchased with the original chamber or added later.

III. Monitoring, Recording, and Alarm System

To ascertain the conditions in the growth chamber, a monitoring and alarm system is required. The monitoring system shall scan and record the temperature, relative humidity, and light intensity, and it shall display these three quantities on charts, dials, or numerical meters so the attendant may observe their levels. The alarm system shall have two set points (high and low) for each of the above variables and shall operate an alarm bell and a manually reset pilot light.

IV. Controls and Other Equipment

Reliability of performance of the growth chamber is of the utmost importance for the successful completion of the research. Therefore, all controls and other equipment used in the growth chamber must be of high quality to insure their reliable operation. They must have zero drift from the set point and a minimum of breakdowns and malfunctions. If a malfunction or breakdown occurs it is important that the trouble be repaired in a short time so as not to destroy the research in progress. Therefore, the controls, relays, switches, breakers, lamp ballasts, motors, and other such items shall be located behind hinged panels that can be opened with a manual latch or a simple screw latch. All electrical components shall be plugged in so that they can be easily replaced.

V. Operating Conditions

The environmental growth chamber will be located in a building that is heated and air conditioned to 21°–24°C and has relative humidity of 40 to 50 percent. If the heating or air conditioning should fail, the temperature could fall as low as 0°C or rise as high as 50°C. The growth chambers must operate as outlined in previous sections over this range of ambient temperatures, and over a range of relative humidities from 20 to 80 percent.

VI. Installation

The vendor must supervise the unpacking or uncrating of the growth chamber and any related equipment. Laboratory personnel will be available to assist the vendor. The vendor must assemble the growth chamber and make all hookups to external services, including electric and water service and sewer hookup. These services will be available within 6 meters (20 feet) of the growth chamber area inside the building.

The vendor will check out and demonstrate to laboratory personnel that the growth chamber is working properly. He will spend at least one additional day demonstrating to the laboratory personnel the proper operation and maintenance of the chamber. The growth chamber shall not be considered as installed until all conditions and requirements have been met or satisfied. (If your institution has regular maintenance men, you may find it less costly to use their service for part or all of the installation.)

VII. Standard of Performance and Acceptance of Equipment
(In this section the various items of performance can be listed. For instance, you might require that the chamber operate for 10–30 consecutive days without a breakdown or failure before the unit will be accepted.)

VIII. Tests during Acceptance Period
During the acceptance period the laboratory will put the growth chamber into operation to ascertain whether it is functioning in accordance with the specifications. The ambient air temperature will be varied from 10° to 50°C for all the tests defined in this section.

A. Temperature
A grid of thermocouples on 15-cm spacings will be placed in the empty chamber on a horizontal plane. Temperatures will be measured at various heights within the growing area.

B. Humidity
The relative humidity will be measured at various locations in the growing area. These humidity measurements will be made at intervals of 10 percent over the range of 50 to 90 percent. Relative humidity measurements will be made when the lights are on and as the temperature is adjusted over the range of 5° to 45°C.

C. Light
A light meter and a quantum flux sensor will be used to check the light intensity and PAR level at several locations in the horizontal plane.

D. Airflow
A hot-wire anemometer will be used to check the air velocity in the vertical or horizontal direction at 30-cm intervals throughout the volume of the growth chamber.

E. Monitoring Equipment
The accuracy of the monitoring equipment will be checked

by independent instruments placed next to the monitoring sensors.
F.   Sealing
To make sure that the growth chamber is sealed inject CO_2 into the chamber until the concentration reaches a level 500 ppm above the ambient air; then observe the decay rate. A maximum reduction of 10 ppm of CO_2 per minute at a differential of 500 ppm between the growth chamber and the air around the chamber will be acceptable.
IX.   Guarantee
The contractor shall furnish all maintenance (labor and parts), including the shipping of old and new parts, for a period of one year beginning on the first day of acceptance. Any periods of downtime, beginning 24 hours after the contractor has been notified of a malfunction, shall be added to the guarantee period so that the user will have one full year of guaranteed operation.

References

AIBS Bioinstrumentation Advisory Council: Controlled Environment Enclosure Guidelines, Part I. 1971. *BioScience* 21(17):913–914. Part II. 1971. *BioScience* 21(19):999–1000. Part III. 1971. *BioScience* 21(21):1083–1084. Conclusion. 1971. *BioScience* 21(22):1131.

ASAE. 1977. Installation and maintenance of farm standby electric power. ASAE Recommendation: R364. Pages 490–492 in *Agricultural Engineers Yearbook*. Ed. J. F. Baxter. Amer. Soc. Ag. Eng., St. Joseph, Mich.

ASHRAE. 1974. Chapter 22, Environment control for animals and plants. Pages 22.1–22.20 in *ASHRAE Handbook of Applications*. Ed. C. W. MacPhee. Amer. Soc. Heating, Refrig., and Air Cond. Eng., New York.

ASHRAE. 1977. Chapter 9, Environmental control for animals and plants. Pages 9.1–9.18 in *ASHRAE Handbook of Fundamentals*. Ed. C. W. MacPhee. Amer. Soc. Heating, Refrig., and Air Cond. Eng., New York.

Bailey, W. A.; H. H. Klueter; D. T. Krizek; and N. W. Stuart. 1970. CO_2 systems for growing plants. *Trans. ASAE* 13(3):263–268.

Campbell, L. E. 1965. Standby electric power equipment for the farm. USDA Leaflet 480. 8 pp.

Canham, A. E. 1966. *Artificial Light in Horticulture*. Centrex, Eindhoven, Netherlands. 212 pp.

Carpenter, G. A., and L. J. Moulsley. 1960. The artificial illumination of environmental control chambers for plant growth. *J. Agr. Eng. Res.* 5(3):283–306.

Chouard, P., and N. de Bilderling, eds. 1972. *Phytotronique II*. Proceedings of symposium, Tel Aviv, March 25, 1970. Gauthier-Villars Editeur, Paris.

Downs, R. J. 1975. *Controlled Environments for Plant Research*. Columbia University Press, New York and London. 175 pp.

Hendricks, S. B., and H. A. Borthwick. 1963. Chapter 14, Control of plant growth by light. Pages 233–261 in *Environmental Control of Plant Growth.* Ed. L. T. Evans. Academic Press, New York.

Morse, R. N., and L. T. Evans. 1962. Design and development of CERES. *J. Agr. Eng. Res.*, 7:(2)128–140.

Pescod, D.; W. R. W. Read; and D. W. Cunliffe. 1962. Artificially lit plant growth cabinets. Pages 175–195 in Proceedings of a Symposium on Engineering Aspects of Environment Control for Plant Growth. Commonwealth Scientific and Industrial Research Organization, Melbourne, Australia.

Chapter **11** DOUGLAS P. ORMROD AND
RICHARD H. HODGSON

Special Purpose Chambers

Many adaptations can be made on growth chambers to meet the special needs of a particular research project. It is not the purpose of this chapter to discuss the many special uses to which ordinary growth chambers have been put, but rather to discuss special chambers and some experiments requiring extensive modification of growth chamber design. Growth chambers used in air pollution, pesticide, and tissue culture research, and in radioisotope studies are presented as examples.

Air Pollution Research

Plant responses to air pollutants can be markedly affected by such environmental factors as light, temperature, humidity, wind velocity, nutrition, soil water status, and the presence of pesticides or other pollutants, including heavy metals, before, during, and after exposure to the air pollutant (Heck 1968; Heggestad and Heck 1971, Hitchcock et al. 1971, Ormrod et al. 1976, Treshow 1970). Growth chamber design should take into account the need to modify these variables and consider the damage that air pollutants can do to growth chamber components as well as to plants.

The effects of air pollutants on plants are studied in controlled environment facilities, in greenhouses, in field exposure facilities, and under open field conditions. Controlled and well-defined environments are recommended for studying pollutant uptake rates, visible and hidden injury, mechanisms of injury development, and amelioration of pollutant effects.

Growth chamber facilities must be capable of monitoring and controlling several factors that are known to affect plant response to air pollutants. Some of these factors are concentration of the pollutant during exposure, length of exposure time, duration and intensity of environmental factors during exposure in relation to the plant phenological condition, and cultural history (Wood et al. 1973). Sophisticated chambers for exposure of a wide variety of plants to air pollutants have been described (Adams 1961, Hill 1967, Menser and Heggestad 1964, Wood et al. 1973). Such chambers provide precise control of temperature, humidity, and light.

The principal components of a facility for air pollutant exposure are a pollutant generator, a filter system, and an exposure chamber, together with a lighting system and a source of temperature- and humidity-controlled air. Chamber design and operation should permit exposure of plants to air pollutants under a range of well-defined environmental conditions at and for particular times.

Most plant-injuring air pollutants react with metal chamber surfaces, often causing corrosion. Chamber surfaces may also absorb the pollutants and even remove them from the atmosphere. Desired concentrations of pollutants in the chambers may be difficult to establish and long equilibration times may be required if any reactive surfaces are present. The ideal exposure chamber would therefore be constructed entirely of materials not affected by the pollutant. Such construction, however, is presently possible only for a single-passthrough air circulation system. For many purposes a recirculating system is preferable, in which case corrosion of metal parts in the system and surface absorption of pollutants must be minimized.

In the single-passthrough system, the air pollutant is injected into the air stream being blown into a glass- or plastic-walled growth chamber. The pollutant does not come into contact with any metal or reactive surfaces. The system must be completely sealed to avoid contamination of the working area. The pollutant is distributed evenly in the air stream by turbulent mixing, by introduction just past a fan, or by use of a large mixing chamber prior to the treatment chamber. The pollutant-air mixture can be distributed throughout the growth chamber by means of perforated manifolds that ultimately exhaust the pollutant outdoors or into an absorbent filter. The plants may be placed on a rotating turntable during the exposure period to ensure uniform average exposure to the pollutant.

If temperature and humidity are to be controlled adequately in a single-passthrough system, the control equipment and pollutant generators must have a large capacity. Moderate to high levels of humidity may be essential for maximal stomatal opening and plant injury.

To ensure precise control of pollutants, independent of outside or laboratory air, designs for single-passthrough systems should incorporate filters to remove pollutants of all kinds from incoming air before the desired pollutants are introduced. Filtration equipment on the air intake should consist of the following sequence: rough (furnace-type) dust filters, high efficiency particulate filters to retain particles above 5 microns, and activated charcoal filters. Air may be drawn initially through a water scrubber that cools and humidifies it and reduces the pollutant content.

As an alternative to the use of remote air conditioning units, the plastic or glass exposure chambers used in a single-passthrough system may be placed within a conventional growth chamber with filtered air drawn from the growth chamber, pollutant added in the incoming duct of the exposure chamber, and pollutant-containing air exhausted through ductwork to the outside (Heck et al. 1968). The environmental conditions in the small chamber will be a function of those in the large chamber, but may not be exactly the same. Differences will depend on airflow rate, construction materials, and radiation characteristics.

Care should be taken during the construction of plastic or glass treatment chambers to use materials that do not themselves emit atmospheric pollutants (see Chapter 5).

The alternative to the single-passthrough system is a recirculating system. This system is particularly useful for experiments on the effect of pollutants on rates of net carbon dioxide exchange, transpiration, and pollutant uptake (Hill 1971, Taylor et al. 1965). The system should have an exhaust fan connected to the ductwork to create a small negative pressure to prevent contamination of the laboratory or work space by any slight leakage of pollutant. There is also the possibility that undesirable gases may accumulate in the chamber atmosphere. Plants may give off trace amounts of terpenes and other volatiles that in turn may affect physiological processes. Some fresh air exchanges are thus required, and a system to control the entry and composition of makeup air at a predetermined rate has to be incorporated in the design. Filtered air should be used for aerated culture solutions and allowance should be made for exhausting it. Care must be taken to exhaust air from the chamber to a safe discharge point. The pollutants must be uniformly distributed within the chamber. In a recirculating system the pollutant should be introduced through a multiport manifold on the suction side of the circulating fans to ensure thorough mixing with the air stream prior to treatment of plants.

Commercial growth chambers of conventional design may be used for a recirculating system but many modifications will be required (Wood et al. 1973). Conventional growth chambers are normally constructed with steel (usually galvanized with zinc), aluminum, copper, and brass parts. Most of the air pollutants studied experimentally, such as sulfur dioxide, ozone, and fluoride, can be very corrosive to these metals. If a conventional growth chamber is used for pollutant studies it may be necessary to replace components or even the entire chamber after extended use. A serious additional problem is that pollutant gases may be absorbed and later released by the materials used in chamber construction. This phenomenon will interfere with any studies of pollutant uptake by plants in the chamber.

Modifications to growth chambers to allow pollutant research should minimize the exposure of metal surfaces and maximize the use of nonmetallic components in the air circulation system. All metal parts, including fans, should be coated with epoxy or Teflon paints. Stainless steel should be used wherever possible, in humidifiers and controls, wall surfaces, cooling coil fans, and tubes. Air ductwork can be constructed of nonmetallic materials although it should be noted that some types of polyvinylchloride (PVC) can be oxidized by pollutants such as ozone and peroxyacetyl nitrate (PAN). Fiberglass can be used for subfloors and perforated floors and walls. Chamber walls can be coated with aluminized plastic or epoxy white paint to maximize reflectivity. All such modifications will increase the cost of the growth chamber facility.

Pesticide Research

Controlled environment facilities are used to evaluate the efficacy of candidate pesticides on a wide range of plant species. Comparisons with known standards can be made quickly, and potential use patterns and precautions identified. Controlled environments are used extensively in fundamental research to determine the absorption, movement, and metabolism of pesticides in plants, and pesticide fate in the environment.

The efficacy and selectivity of a pesticide may depend on the plant or animal species involved, its stage of development, and the environmental conditions prevailing before, during, and after its exposure (Muzik 1976). The degradation and fate of a pesticide and the chemical nature, amount, and bioavailability of residues may be altered significantly by environmental factors (Brian 1970). An understanding of environment-plant-pesticide inter-actions and a definition of environmental parameter effects are, therefore, essential and can be gained through the use of appro-priate research methods (Hammerton 1968).

Air and soil humidity affect the uptake, movement, and activ-ity of pesticides. Light intensity, temperature, and wind velocity are other important factors that can affect their activity. Absorp-tion and retention of pesticides by plants and their toxicity to plants are influenced by the size, shape, and thickness of leaves, the amount of cuticle on plant surfaces, and in some cases, on the depth and ramification of the root system; all of which are affected by environmental conditions during growth before pesti-cide application. Environmental conditions at the time of treat-ment can influence pesticide performance or efficacy by deter-mining the rate of volatilization of leaf and soil deposits, the rate of desiccation of pesticide deposits on leaves, and the degree of absorption and movement of soil-applied pesticides. Environ-mental conditions after treatment can affect the persistance and distribution of residues on leaves and in the soil. These conditions also affect the rate and degree of reactivity toward insects, patho-gens, and weeds, as well as the desired plants, thus determining whether insect, pathogen, host, weed, or crop will recover from pesticide treatment.

Pesticide researchers use growth chambers as a reproducible environment to provide plant material that may be used intact, as plant parts, or as cellular or subcellular material for short term experiments carried on outside the chamber (Blankendaal et al. 1972). Growth chambers also provide a well-defined constant or reproducible environment in which varied pesticide treatments are made (Eshel and Katan 1972, Van Oorschot 1970). Growth chambers permit experiments in which environment is a variable and generally superimposed on varied pesticide treatments (e.g. Hodgson 1971, Hodgson et al. 1974). For such experiments, the requirements for controlling and manipulating the environment are the same as those for many other physiological and hor-ticultural experiments.

The physical, chemical, and physiological properties of pesticides all must be taken into account when they are used in controlled environment research. Many pesticides and their metabolites are volatile enough to be lost to the growth chamber atmosphere in significant quantities from treated soil, nutrient solutions, or plant surfaces within a few hours or days (Van Valkenburg 1973, Weed Sci. Soc. Amer. 1974). Once released to the chamber atmosphere these compounds will be recirculated within the chamber, or carried into adjoining spaces with the exhaust air stream. Whether they escape from a particular growth chamber or not, they may adsorb on initially untreated plant surfaces and be absorbed, or they may adsorb on plant containers or chamber components and subsequently be rereleased. Thus, the pesticide treatment can be different from that intended, the results ambiguous, apparatus and neighbouring experiments contaminated, and hazardous materials may be released. Provision should be made for scrubbing the air entering the growth chambers and venting, selectively filtering, or scrubbing the exhaust gases. If a common venting system is used for several chambers, care should be taken to ensure that exhaust gases from one cannot backflow into another.

Most pesticide absorption, translocation, and metabolism research with plants requires the use of radioactively labeled compounds—pesticides, their metabolites, or formulating agents. Some special advantages and precautions result from the use of radioactively labeled pesticides in growth chambers. One advantage is that a balance sheet can be developed for the disposition of a labeled pesticide and its metabolites. The degree of adsorption to container walls and other apparatus can be measured, vapor losses can be monitored, and overall recovery can be measured. Indeed, such studies often yield more sensitive measures of potential contamination problems than do experiments with unlabeled materials where assessment of contamination is limited to standard chemical and bioassay techniques. As a result, investigators often take more care in designing tracer experiments with decontamination and disposal in mind than they do for studies with nonradioactive pesticides. In fact, most of the potential contamination problems are similar in both cases.

Pesticide research is often expedited by use of a chamber within a chamber, with a single-passthrough conditioned airstream. The advantages of such internal chambers are that they can be constructed entirely of inert materials, they can assist in confining the pesticide to reduce or eliminate growth chamber contamination, and they can be removed to special facilities for decontamination.

Growth chambers for pesticide research should be isolated physically from other research facilities, either in a separate building or in a section of building having isolated ventilation systems and physical barriers to diffusion and forced movement of volatile materials. Entry from other areas should be by means of double doors and air locks. The entire pesticide research section should be maintained under slight vacuum by fans that exhaust to the outside of the building through a stack with appropriate filters and wash-down capability to remove pesticides from the air system. Individual growth chambers should also have slight negative pressure compared to the surrounding work area to ensure that leakage is only inward.

Tissue Culture Research

Plant cell and tissue cultures are used to characterize normal and diseased growth and differentiation in plants (Hildebrandt 1973); to determine growth regulator interactions (Epstein et al. 1975, Stearns and Morton 1975); to study biotransformations (Steck and Constabel 1974) and metabolite (Kehr 1975) or enzyme production (Ray and Still 1975); and for long term tissue storage. In organ culture, excised roots, shoots, leaves, flowers, or fruits are grown under aseptic conditions. Growth or metabolism of these organs may be studied, their life cycle completed to obtain seeds, or isolated embryos may be grown and studied. In any case, the cells and tissues are isolated and grown under controlled environments on a synthetic medium.

As for other plant research, the use of controlled environment chambers for tissue culture research is essential for repeatability and comparability between locations. In addition, controlled environment facilities are needed because successful culture requires that some environmental factors be held within narrow ranges. The environmental requirements for many tissue cultures are not well defined, and growth chambers are used for systematic examination of environmental effects.

A constant temperature in the range of 25°–30°C has been standard for tissue culture. This temperature is not consistent, in many cases, with the conditions that exist in the natural habitats of some species; other temperature ranges may be more satis-factory for particular species. Some species require temperatures that fluctuate diurnally and change to simulate seasons.

Light is not required for multiplication of plant cells as callus or free suspensions and may retard growth (Murashige 1973). However, many cell cultures intended for organ initiation need light and, in most cases, experimental work must yet be done to determine the optimum intensity, duration, and quality. Stem tip or embryo cultures often grow best in the light, while root tip cultures grow well in light or dark (Hildebrandt 1973). Chlo-rophyll and other pigments may develop in cultures grown in light and the quality and quantity of light may influence the amount of pigment formation (Vasil and Hildebrandt 1966). Light quality and quantity will also influence the development of roots, stems, leaves, and plants from undifferentiated cells.

Maintenance of at least 35 to 50 percent relative humidity in the growth facility is important to reduce excessive water loss from cultures (Hildebrandt 1973). Loss of water affects the physi-cal qualities of semisolid media and increases the effective con-centration of solutes in both semisolid and liquid media. Al-though water loss from cultures cannot be prevented easily, if gas exchange is to be permitted it must occur reproducibly with time to ensure uniformity of culture metabolism, growth, and development.

Radioisotope Research

Special growth chambers may be used for the production of plant material labeled with carbon-fourteen. Several such ^{14}C bio-synthesis chambers have been described (Anderson et al. 1961, Jenkinson 1960, Smith et al. 1962). Major modification of the chamber is required to allow plants to be grown for a long period in an atmosphere containing carbon dioxide labeled with [14]C. The chamber must be entirely sealed, with pressure-tight electrical fittings and refrigerant tubing.

The use of hydroponic culture will simplify the operation. Soil should not be used because of CO_2 release from the soil and be-cause of the difficulty of removing roots at harvest. Use of hy-droponics also simplifies watering of plants. One method of water-ing in a sealed growth chamber is to return condensate from the cold air conditioning coil to the pots, after passing it through an ion exchanger to remove the heavy metals.

The most difficult problem in a sealed ^{14}C biosynthesis chamber is the maintenance of CO_2 levels. One satisfactory method is to use labeled barium carbonate solution to achieve the desired specific activity of ^{14}C. Addition of acid to the labeled bicarbonate solution mixture releases $^{14}CO_2$. A switch on a CO_2 analyzer recorder opens solenoid valves that allow separate entry of acid and bicarbonate solution into a reaction vessel in the sealed chamber.

A safety valve on the sealed chamber is needed in case of substantial pressure fluctuations. The valve should be attached to a series of traps containing sodium hydroxide solution that washes the air free of CO_2 and radioactivity.

Growth chambers are used for tracer studies with labeled pesticides, growth regulators, metabolites, or inorganic salts. Many of these materials used in tracer studies are labeled with ^{14}C. If the experimental material is degraded during the experiment and the labeled atoms are oxidized and released as $^{14}CO_2$, special problems may arise. Significant amounts of $^{14}CO_2$ may be reabsorbed by the plant tissue in chambers in which air is recirculated, giving rise to labeled compounds other than those directly connected with the metabolism of the originally applied compound. Unexplained loss of radioactivity may occur in leaky recirculating chambers, or in unmonitored single passthrough chambers, and plants in neighbouring spaces may acquire unanticipated burdens of radioactive plant metabolites. Similar problems could arise from loss of tritiated water following degradation of compounds labeled with ^3H.

In all experiments with radioactively labeled compounds consideration must be given to the following questions: Is the chamber uncontaminated before use? Are the compound or its metabolites volatile? Are any volatile products phytotoxic or toxic to animals? Are all radioactive materials accounted for so that a balance sheet can be developed for the experiment? Will it be possible to decontaminate the chamber and other associated hardware when the experiment is over even though many compounds are absorbed to surfaces very tenaciously? The answers to these questions will indicate whether such experiments should be conducted in the type of sealed chamber described for growing plants labeled with ^{14}C.

References Cited

Adams, D. F. 1961. An air pollution phytotron: A controlled environment facility for studies into the effects of air pollutants on vegetation. *J. Air Pollut. Control Ass.* 11:470–476.

Anderson, A.; G. Nielsen; and H. Sorenson. 1961. Growth chamber for labeling plant material uniformly with radiocarbon. *Plant Physiol.* 14:378–383.

Blankendaal, M.; R. H. Hodgson; D. G. Davis; R. A. Hoerauf; and R. H. Shimabukuro. 1972. Growing plants without soil for experimental use. Ag. Res. Serv., USDA Misc. Publ. 1251. 17 pp.

Brian, R. C. 1970. Environment and herbicide activity. *Agr. Prog.* 45:48–57.

Epstein, E.; I. Klein; and S. Lavee. 1975. Uptake and fate of IAA in apple callus tissues: Metabolism of IAA-2-^{14}C. *Plant Cell Physiol.* 16:305–311.

Eshel, Y., and J. Katan. 1972. Effect of dinitroanilines on solanaceous vegetables and soil fungi. *Weed Science* 20:243–246.

Hammerton, J. L. 1968. The environment and herbicide performance. *Proc. 9th. British Weed Control Conf.* 1088–1110.

Heck, W. W. 1968. Factors influencing expression of oxidant damage to plants. *Ann. Rev. Phytopath.* 6:165–188.

Heck, W. W.; J. A. Dunning; and H. Johnson. 1968. Design of a simple plant exposure chamber. Nat. Center Air Pollut. Contr. Publ. APTD-68-6. 24 pp.

Heggestad, H. E., and W. W. Heck. 1971. Nature, extent and variation of plant response to air pollutants. *Advan. Agron.* 23:111–145.

Hildebrandt, A. C. 1973. Plant cell suspension culture. Pages 215–219 in P. F. Kruse, Jr., and M. K. Patterson, Jr., eds., *Tissue Culture Methods and Applications.* Academic Press, New York.

Hill, A. C. 1967. A special purpose plant environmental chamber for air pollution studies. *J. Air Pollut. Control Ass.* 17:743–748.

Hill, A. C. 1971. Vegetation: A sink for atmospheric pollutants. *J. Air Pollut. Control Ass.* 21:341–346.

Hitchcock, A. E.; D. C. McCune; L. H. Weinstein; D. C. MacLean; J. S. Jacobson; and R. H. Mandl. 1971. Effects of hydrogen fluoride fumigation of alfalfa and orchard grass: A summary of experiments from 1952 through 1965. *Contrib. Boyce Thompson Inst.* 24:363–386.

Hodgson, R. H. 1971. Influence of environment on metabolism of propanil in rice. *Weed Science* 19:501–507.

Hodgson, R. H.; K. E. Dusbabek; and B. L. Hoffer. 1974. Diphenamid metabolism in tomato: Time course of an ozone fumigation effect. *Weed Science* 22:205–210.

Jenkinson, D. S. 1960. The production of ryegrass labeled with carbon-14. *Plant Soil* 13:279–290.

Kehr, A. E. 1975. New developments in plant cell and tissue culture. *HortScience* 10:4–5.

Menser, H. A., and H. E. Heggestad. 1964. A facility for ozone fumigation of plant materials. *Crop Science* 4:103–105.

Murashige, T. 1973. Somatic plant cells. Pages 170–172 in P. F. Kruse, Jr., and M. K. Patterson, Jr., eds., *Tissue Culture Methods and Applications.* Academic Press, New York.

Muzik, T. J. 1976. Influence of environment on response of plants to herbicides. Pages 203–207 in vol. 2, L. J. Audus, ed., *Herbicides: Physiology, Biochemistry, Ecology.* 2d. ed. Academic Press, London.

Ormrod, D. P.; N. O. Adedipe; and D. J. Ballantyne. 1976. Air pollution injury to horticultural plants: A review. *Hort. Abstr.* 46:241–248.

Ray, T. G., and C. C. Still. 1975. Propanil metabolism in rice: A comparison of propanil amidase activities in rice plants and callus cultures. *Pestic. Biochem. and Physiol.* 5:171–177.

Smith, J. H.; F. E. Allison; and J. F. Mullins. 1962. Design and operation of a carbon-14 biosynthesis chamber. Ag. Res. Serv., USDA Misc. Pub. 911. 15 pp.

Stearns, E. M., Jr., and W. T. Morton. 1975. Effects of growth regulators on fatty acids of soybean suspension cultures. *Phytochemistry* 14:619–622.

Steck, W., and F. Constabel. 1974. Biotransformations in plant cell culture. *Lloydia* 37:185–191.

Taylor, O. C.; E. A. Cardiff; and J. D. Mersereau. 1965. Apparent photosynthesis as a measure of air pollution damage. *J. Air Pollut. Control Ass.* 15:171–173.

Treshow, M. 1970. *Environment and Plant Response.* McGraw-Hill, New York. 422 pp.

Van Oorschot, J. L. P. 1970. Influence of herbicides on photosynthetic activity and transpiration rate of intact plants. *Pestic. Sci.* 1:33–37.

Van Valkenburg, W. 1973. *Pesticide Formulations.* Marcel Dekker, New York. 481 pp.

Vasil, I. K., and A. C. Hildebrandt. 1966. Growth and chlorophyll production in plant callus tissues grown *in vitro. Planta* 68:69–82.

Weed Science Society of America. 1974. *Herbicide Handbook,* 3d ed. Weed Sci. Soc. Amer., Champaign, Ill. 430 pp.

Wood, F. A.; D. B. Drummond; R. G. Wilhour; and D. D. Davis. 1973. An exposure chamber for studying the effects of air pollutants on plants. Penn. State Univ. Ag. Exp. Sta. Rept. 335. 7 pp.

Chapter **12**   ROBERT W. LANGHANS AND P. ALLEN HAMMER

Chamber Maintenance

Growth chamber maintenance may be considered a large liability. Few items in a department supplies and expenses budget are as large as the cost of a well-maintained growth chamber; if not, then the conditions in the chamber are likely to be poor. The very nature of the design, the number of lamps, heating elements, cooling coils, mist and steam jets, and fans—all working to maintain high light intensities and close temperature and relative humidity tolerances—indicate hard work. Because of the precision required, the chamber equipment is always running, usually in an "on/off" mode, which is the hardest kind of operation. To keep everything working properly one must plan an efficient maintenance program.

Large growth chamber installations should be able to afford personnel who only perform maintenance on the chambers. As a rule of thumb, one person can care for about 25 chambers, making daily observations, checking the controls, troubleshooting, and upgrading and replacing worn-out equipment.

Regular Maintenance

A regular maintenance program is superior to trouble shooting maintenance. A regular program will prevent most breakdowns, whereas trouble shooting only occurs after a breakdown. One of the real limitations of a plant growth chamber is the possibility of a breakdown. Even a few hours of "down" time can be sufficient to destroy an experiment. Breakdowns are very expensive, although it is sometimes hard to put a dollar value on the damage. Whenever possible a daily inspection of equipment should be

Table 12.1. Chamber maintenance checklist

DAILY

Check recording instruments for temperature, RH, and light conditions.
Observe lamps and thermometers if recording instruments are not used.

WEEKLY

Check condensing unit for compressor oil level, refrigerant level and unusual noises.
Inspect all fan motors.
Inspect chamber drains.

QUARTERLY

Clean light cap filters and all other filters.
Clean air-cooled condenser.
Check all belts for wear, aging, and slipping.

ANNUALLY

Inspect and clean control panels, drawers, and consoles.
Check operation of switches, relays, magnetic starters, and related electrical
    components.
Tighten all electrical connections.
Check current draw on major electrical components, i.e. heaters, compressors, and
    lamps.
Calibrate all controllers and recorders.
Inspect chamber hardware, door hinges, rollers, and latches, and lubricate if
    required.
Inspect condensing unit, related piping, and fittings for oil and refrigerant leaks.
Change compressor oil.

AS REQUIRED

Lubricate all motors and fans according to manufacturer's specs.
Replace chamber lamps.
Clean light cap barriers.

Compiled by Gerald Bergenstock, Growth Chamber Maintenance Supervisor,
Department of Plant Pathology, Cornell University.

made by a qualified maintenance supervisor. An inspection will probably catch most potential problems before they develop and prevent the stoppage, the lost experiment, or the complete destruction of a piece of equipment.

Fluorescent Lamps

The most frequently replaced items are the lamps. The major input wattage in the typical chamber is provided by fluorescent lamps. Lamp manufacturers suggest 5000 or more hours of useful life, but in our experience it has been about 2500 hours. High temperatures cause a rapid deterioration of the lamps. The common cool white or warm white fluorescent lamp has its greatest life in ambient air temperatures around 25°C. Many chambers, because of the proximity of the lamps, the small enclosed area, and the lack of air conditioned air, are not able to maintain this temperature. High temperatures prevail and shorten lamp life. Fluorescent lamp life can be measured by the decrease in irradiance. We have recorded losses of as much as 100 ft-c per week. If irradiance is important, then the lamps should be changed frequently to maintain a steady level. For example, changing one quarter of the lamps each month will achieve a nearly constant light level.

The ease of lamp change should be considered when purchasing a chamber. The chambers should not have to be emptied of plants and benches to allow for lamp changes.

The average chamber runs on a sixteen-hour photoperiod, or approximately 5000 hours of light per year. If irradiance is not critical in the experiment, one change per year will be sufficient. If irradiance is critical, more frequent changes will have to be made.

Most chambers use about 3 lamps per linear foot of chamber. Estimated costs are $2.50 per square foot of floor area per year with one change or $5.00 per square foot of floor area per year for two changes of lamps.

Burned out lamps are generally not a problem if the lamps are changed annually. The lamps are brightest during the first 100 hours, then the intensity drops sharply before assuming a steady rate of decline. The experimenter may wish to burn the lamps for a week or so before critical experiments are started. Fluctuating temperatures will also affect the output of the lamps.

Incandescent Lamps

In most growth chambers 10 to 30 percent of the input wattage of light is provided by incandescent lamps. These lamps should be changed when the fluorescent lamps are changed. Since most of the lamps are low wattage (40 to 60 watts), their life is shorter than the fluorescent. Long-life incandescent lamps are more expensive but they should be considered. Lamps such as those used for traffic signal lighting are designed for longer life. Many incandescent lamps burn out between normal changes. They should be checked once a week.

HID Lamps

Several types of high intensity discharge (HID) lamps are now being used in growth chambers; metal halide, high pressure sodium, and mercury. These lamps are expensive, but last 15,000 hours or more. A chamber of mercury lamps is capable of producing irradiance 50 to 100 percent brighter than the conventional fluorescent chambers. They are more efficient and produce less heat. The lamps are not particularly affected by temperature. HID lamps have not been used in growth chambers for very long, and researchers have little experience with their problems.

Barriers

Many chambers have some type of barrier between the light cap and the growing area. Dirt and dust are carried with the cooling air and settle on the barrier. Under normal conditions it should be cleaned at least every three months. Since most barriers are made of plastic (normally plexiglass), abrasive cleansers should not be used. Scratches interfere with and reduce light transmission. After a few years the plastic becomes brittle and may crack.

Heating Equipment

Electric heating elements do not often go out of adjustment; either they work very well or they are burned out and do not work at all. Malfunctions, however, may not be readily noted unless the temperatures are carefully monitored. The clue to a malfunction would be excessive cycling of the temperature. It is easy to feel whether warm air is coming from each element during a heating cycle. Looking at the heating elements and checking with an ohmmeter is also possible, but the elements are usually buttoned up deep in the walls and not easily visible. Because of the high humidity in the chambers, the elements do corrode and short out. An inspection should be made at least once a month.

Cooling Equipment

The cooling coils are usually trouble free except for the debris, dead leaves and dust, that accumulates in and on the coils. If the coils are cleaned every 6 months there should be little difficulty.

The compressor on most chambers can be expected to last for 5 years. Weekly inspection of the belts, compressor oil level, and refrigerant level must be made when the equipment is running. The compressor is probably the most vulnerable piece of equipment in the chamber and the most expensive to replace. Breakdowns in the cooling system are usually related to the compressor. Heat and vibration are the major sources of trouble. As a rule a compressor costs $150 per horsepower.

Thermostats

Thermostats, like any other piece of equipment, fall out of adjustment, get dirty, and break. Usually a breakdown is obvious and the thermostat must be replaced. When a thermostat is out of adjustment or dirty, its sensitivity is reduced and the problem is more difficult to detect. We suggest frequent monitoring of the air temperature with a precise temperature recorder, such as a thermocouple. Avoid using the setting on the thermostat, as we have seen many "accurate" thermostats off by 5°C. We suggest taping the dial so the numbers cannot be read. At least every 6 months the thermostat should be inspected and cleaned.

Some electrical thermostats have open contacts and they will fail in a few years under the high relative humidity of most growth chambers. If the thermostat has open contacts, replacement every 2 years is recommended, whereas closed electrical contact or pneumatic thermostats will last 10 years without replacement.

Humidification Equipment

Steam and misted water in the air stream are the most common methods of increasing the relative humidity. Both systems require regular maintenance to be sure the orifices of the steam or mist nozzles are kept open and free of salt. Solenoid valves should be inspected regularly for proper operation.

Cleanliness of the element controlling the relative humidity is a major factor since a great deal of air passes over the element. The element is usually damp so that dust sticks and accumulates, making the controls unreliable. When relative humidity control is important the humidity level should be checked daily with a dewpoint thermometer or psychrometer and the controls adjusted. The control element should be cleaned at the start of each experiment.

Ventilation Systems

Fans. Air in the chamber is kept in motion by some type of fan. This air movement is most necessary for accurate temperature and humidity control. Most installations use oilless motors so oiling is unnecessary. Motors, however, burn out in moist conditions in chambers. One should expect 5 years of life from the average fan motor. Extra fan motors should be on hand to replace burned-out equipment immediately as each fan is very important for proper distribution of air in the chamber.

Louvers. Many chambers have adjustable louvers at the air outlets and the most advantageous position should be used. This position can be determined from the manufacturer's suggestions or by experimentation.

Filters. The makeup air for the chamber and light cap is usually filtered. The filters are similar to those found on household furnaces. These filters should be inspected once a month and cleaned and replaced as needed. A dirty filter reduces the amount of incoming air and makes the fans work harder, causing premature wear and shortening their useful life, as well as preventing the proper circulation of air for accurate temperature control.

Walls

The walls of the growth chambers are white or mirrorlike to reflect and maximize the light in the chamber. As water and other solutions are splashed and spilled, the walls get stained and dirty. The reflectance of the wall should be inspected at least once a year. If the walls are dirty they should be cleaned or recovered.

Breakdowns and Emergencies

Fail safe. The objective of a fail safe control is to prevent the growth chamber from destroying itself. This is usually done by high and low temperature limit switches, which when activated turn off the growth chamber and sound an alarm. The switches should be set at reasonable levels within a few degrees of the tolerances of the experiment. These controls are infrequently used and should be inspected for proper setting and operation every few months.

Back up equipment. Spare modular electronics controls, temperature and relative humidity sensors, fan belts, and a motor for each type of fan in the growth chamber should be on hand for immediate replacement. A spare compressor with motor is also desirable and could avoid a prolonged delay.

Growth chambers are completely dependent on electricity, and a back-up generator might seem to be necessary. In our opinion, however, the size and cost of a generator that would be used only to run the chambers, are out of proportion to its worth. If standby generators are available, however, plans should be made to use them in case of power failure.

Water cooling systems. The water cooling system used to remove the heat from the refrigeration system is a potential trouble spot in some chambers. The system may be quite remote from the actual installation. If the system fails the compressor will overheat and stop. An alternate water supply and dual water pumps in the system are suggested as a safeguard.

Chapter 13   P. ALLEN HAMMER AND ROBERT W. LANGHANS

Experimental Design

The experimental design and statistical analysis of field and greenhouse experiments has been adequately discussed, while little attention has been given to statistical problems unique to growth chamber experiments. Went (1957) reported reduced plant phenotypic variability in growth chambers compared to greenhouses and from these data he proposed that less plant replication was needed in growth chamber experiments. This is perhaps a dangerous assumption; we will explore this recommendation and provide insight into some problems in the design and analysis of growth chamber experiments.

Growth chambers have been used to provide the researcher with controlled environmental conditions in which to grow plants. They are used for two general purposes. First, the chamber is used to produce a standard or defined environment for the growth of plants (commonly very important for the plant breeder or plant physiologist). Second, the chamber is used to study the effect of one or more controllable environmental parameters on growth (commonly very important for the horticulturist and agronomist).

Common Experimental Problems

There are many causes of environmental variability in growth chambers (Hammer and Langhans 1972). They include (1) light decay over time, (2) changes in irradiance with distance from the source, (3) spatial distribution of light, (4) variation in temperature within a chamber, and (5) differential air movement. Many of these causes have also been reported by other authors (Carlson,

Motter, and Sprague 1964, Gentner 1967, Measures, Weinberger, and Baer 1973) and they certainly are important regardless of the purpose of the research being conducted in the growth chamber.

If growth chambers are to be used for maximum experimental precision, the investigator must be aware of the causes of environmental variability and use experimental designs to account for them. One of the best and simplest designs is a randomized complete-block design, with square or nearly square blocks (Snedecor and Cochran 1971), which significantly increases precision over a completely random design with the same number of plants. However, if limited space makes it difficult to use more complex experimental designs, the experimenter should at the very least randomize the plants within a chamber. Experimental error will be greater, but unwanted variation will be equally distributed over the plant material.

The use of guard rows will prevent an edge or border effect. Although no measurements of the edge effect have been made in the growth chamber, guard or border rows are often included in greenhouse and field experiments and they are certainly appropriate for growth chamber experiments. This practice is particularly important when large plants that interfere with air movement and light distribution are being grown.

Standard Conditions

When the growth chamber is used as a standard environment in which to grow plants, a given set of environmental parameters should be maintained. Problems such as decay of light, increase in plant size, and drift in temperature settings can be very troublesome in long-term studies. Accurate measurements of the environmental conditions are necessary, in order to compare results from experiments conducted at different times or in different chambers set at the "same" environmental conditions. This measure of "sameness" should be precise enough to account for any environmental parameter that may interact with the applied treatments. Suggested methods of measurement are given in the appropriate sections of this handbook.

When experiments are repeated, control plants of the same species (nontreated plants) or a standard plant different from those being studied can be used to monitor or confirm the environmental parameters. For example, if the researcher knows or suspects that a small difference in air temperature may affect the results of an applied treatment, then the standard plant should accurately measure a small difference in air temperature, if it is to be used as a basis for comparison. The standard plant method is the simplest (requiring the least amount of equipment) and most accurate method when properly used.

Variable Environmental Conditions

Growth chambers provide an exciting research tool for studying the effects of the environment on plant growth and development. Environmental parameters, when properly controlled and measured, can be programmed and maintained over a wide range of conditions. Not only the effect of a single environmental parameter but the interaction of several environmental parameters can be investigated.

Replication becomes a very difficult problem in the variable environmental studies. Multiple plants in a single chamber under a set of environmental parameters give only an estimate of sampling error. Replication should be done in several chambers or repeated in the same chamber, to estimate experimental error. Of course, these initial experiments may indicate only a small experimental error relative to the sampling error. Our experience suggests this is not the case with most plants. The possibility of variability within a chamber suggests the need for estimating experimental error between chambers and within chambers before accepting multiple samples within a chamber as an estimate of experimental error, when the treatment is an environmental parameter and only one setting of a growth chamber is tested at a time.

Determining Experimental Design

The following steps are considered minimal in determining the experimental design used in a particular growth chamber.

1. Measure all of the environmental parameters (light, temperature, relative humidity, carbon dioxide concentration, and air flow) at multiple points within the chamber. These measurements will indicate the desirable and undesirable locations in the chamber. The variations in temperature and light intensity are especially critical (Figs. 13.1 and 13.2). If environmental variation is evident, block to minimize its effects; group treatments in areas of greatest uniformity within the chamber.

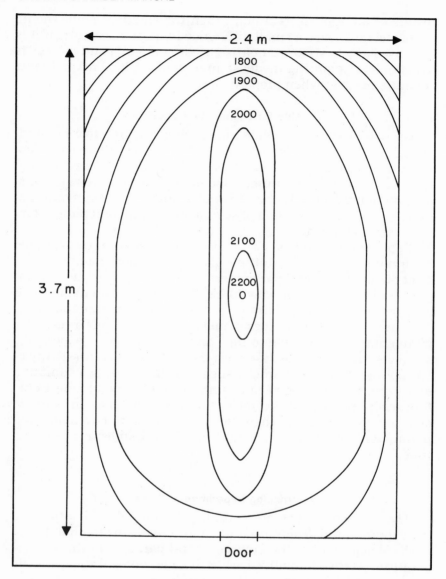

Fig. 13.1. Contour map of light intensities (ft-c) measured 0.91 m below the barrier in Cornell Growth Chamber. The lamps (General Electric warm white fluorescent [F96PG17/WW/VHO] and 40 watt incandescent, with 84% and 16% of the input wattage, respectively) were in excess of 5000 hr of usage. Each contour line represents a decrease in 100 ft-c from the center of the chamber. (Hammer and Langhans 1972.)

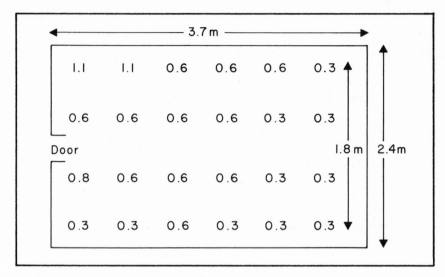

Fig. 13.2. Temperature (±°C) variation within a Cornell growth chamber at 24 locations 0.91 m below the light barrier with setting (23.9°C) constant for all readings.

2. Grow the experimental plant (or a selected standard plant) in the block arrangement (at least three times) within the same chamber using a standard set of environmental parameters. If several chambers are to be used in the experimental work to follow, these initial experiments should be done in all of them.

3. Run an analysis of variance on the data and determine the relative importance of blocks, multiple run variance, between chamber variance, and sampling variance to your proposed experiment. These estimates of variance will help in determining the appropriate experimental design as well as the number of replicates needed for a selected level of precision (Federer 1963).

Most researchers will object to the amount of preliminary experimentation suggested here. Of course, the estimates of variance can be made in the proposed experiment. However, we feel that in most cases (particularly for the new growth chamber researcher) the preliminary experiments provide not only estimates of variance, but also valuable experience in growth chamber experimentation.

Conclusion

Variation is an important factor in growth chamber experimentation. We feel that Went's suggestion of using a few replications may be wrong in many growth chamber experiments. One reason for using growth chambers is the much reduced experimental error, and everything possible should be done to minimize it. Experimental error can be reduced by blocking, and we have also suggested guard rows.

Replication of the treatment in growth chamber studies is an important concern. When an environmental parameter is the treatment, the replication should be made on the treatment. Multiple plants within a chamber provide only sampling error, at least until it is shown that no difference exists between runs or chambers.

Each experiment conducted in a growth chamber should be considered unique and the best statistical approach for that experiment should be used. A statistician should be consulted in the planning stages of each experiment.

References

Carlson, G. E.; G. A. Motter, Jr.; and V. G. Sprague. 1964. Uniformity of light distribution and plant growth in controlled environment chambers. *Agron. J.* 56(2):242–243.

Federer, Walter T. 1963. *Experimental Design, Theory and Application.* Macmillan, New York. 544 pp.

Gentner, W. A. 1967. Maintenance and use of controlled environment chambers. *Weeds* 15(4):312–316.

Hammer, P. Allen, and Robert W. Langhans. 1972. Experimental design considerations for growth chamber studies. *HortScience* 7(5):481–483.

Measures, Mary; Pearl Weinberger; and H. Baer. 1973. Variability of plant growth within controlled-environment chambers as related to temperature and light distribution. *Can. J. Plant Sci.* 53:215–220.

Snedecor, George W., and William G. Cochran. 1971. *Statistical Methods.* Iowa State University Press, Ames.

Went, F. W. 1957. Genotypic and phenotypic variability. Pages 195–201 in *Experimental Control of Plant Growth.* Chronica Botanica, Waltham, Mass.

Chapter **14**

ASHS COMMITTEE ON
GROWTH CHAMBER ENVIRONMENTS*

Guidelines for Reporting Studies in Controlled Environment Chambers

The precise regulation of environmental parameters made possible by growth chambers has permitted investigators to study many plant processes with a precision and reproducibility not available otherwise. However, the environmental conditions employed in growth chamber studies often are not reported in detail sufficient to allow either comparison of the results with similar experiments or repetition of the studies in other laboratories. The following guidelines are presented to assist investigators to achieve these objectives. The guidelines also may alert investigators to factors of the environment that could be important in their experiments but that they do not measure at the present time.

These guidelines were first published in 1972 (*HortScience* 7:239) and have been revised to reflect changes in measurement techniques or instrumentation based on research experience and improvements in measuring devices. For example, we recommend reporting visible radiation in units of photon flux density rather than in photometric units.

Each investigator and publication may prefer to report experimental parameters differently, but it is hoped that the sample text

*Wade L. Berry, University of California, Riverside; P. Allen Hammer, Purdue University; Richard H. Hodgson, Agricultural Research Service, USDA, Fargo, North Dakota; Donald T. Krizek, Agricultural Research Service, USDA, Beltsville, Maryland; Robert W. Langhans, Cornell University; J. Craig McFarlane, Environmental Protection Agency, Las Vegas, Nevada; Douglas P. Ormrod, University of Guelph; Hugh A. Poole, Ohio Agricultural Research and Development Center, Wooster; Theodore W. Tibbitts, University of Wisconsin, Madison.

following the guidelines will be helpful. A tabular presentation of critical parameters may sometimes be preferred to text.

Guidelines

Minimum requirements are numbered; optional measurements are lettered.

Irradiation

1. Lamp types and percent of input wattage for each type.
2. Light readings. Report values preferably in $nE \cdot s^{-1} cm^{-2}$ or $mWcm^{-2}$ for the 400–700 nm wave band.* Give type, spectral sensitivity, make, and model of meter used.
3. Location of meter reading in relation to plant canopy.
4. Photoperiod. Indicate whether lights are turned on gradually or abruptly; if gradual, indicate program. Indicate length of diurnal cycle if other than 24 hours.
5. Lamp barrier. Indicate if present or absent; if present, indicate material used.

   a. Manufacturer and designation of lamps. (a) For incandescent and high intensity discharge (HID) indicate rated wattage and operating voltage; (b) for fluorescent indicate loading, 400 ma, 800 ma, or 1500 ma.

   b. Changes in irradiance in space and over time.

   c. Total radiant energy ($mWcm^{-2}$), indicating instrument and its range of measurement.

   d. Gradient in irradiance over the growing area.

   e. Spectral energy distribution or spectroradiometric curve.

Temperature (°C)

1. Air temperature with a shielded sensor. Indicate type and location of sensor in relation to the plants, and temperature values for day and night.
2. Substrate temperature. Indicate type and location of sensor in the substrate and temperature values for day and night.
3. Thermoperiod. Indicate if day-night or other program is abrupt or gradual.

*Light is reported variously as photosynthetically active radiation (PAR), photosynthetic photon flux density (PPFD), or photosynthetic irradiance (PI). (See R. Shibles, 1976, Terminology pertaining to photosynthesis, *Crop. Sci.* 16:438–439; and Chapter 1, Light.)

a. Gradient of air temperature over the growing area.

b. Soil thermoperiod in relation to photoperiod.

c. Leaf temperature, indicating method and location of measurement.

Relative Humidity (%) or Dew Point (°C)

1. Day and night values. Indicate type and location of sensor.

a. Variations from established level during temperature cycling, and changes occurring during the experiment.

Carbon Dioxide

1. Extent of makeup air. Indicate frequency of complete exchange or rate of addition, giving internal chamber volume.

a. Level in the plant area. Variations during light and dark periods, indicating instrument used to monitor or control CO_2 levels.

Air Movement

1. Direction of movement (up, down, or horizontal).

2. Air flow rates at top of plant canopy.

a. Variations in air flow over the growing area at the beginning and end of the experiment. Indicate measuring instrument used.

Container, Substrate, and Nutrients

1. Composition, capacity, shape, and color of container.

2. Substrate.

3. Nutrient solution. Indicate macronutrient concentration in meq per liter and micronutrients in ppm, frequency and volume of addition, and chelates used.

4. Method of solution renewal or replacement in liquid cultures.

a. Concentration of nutrients at time of solution renewal or replacement, and at the end of the study in liquid culture experiments.

b. Substrate solution pH and pH fluctuations.

c. Tissue analysis.

d. Source of substrate.

Sample Text

Studies were conducted in a 3 m³ reach-in chamber fitted with a Transpex* barrier and having 75% input wattage of 1500-ma cool white fluorescent and 25% input wattage of Lumo 100-W 130-V extended-service incandescent lighting. The irradiance at the top of the plants was 32.5 ± 1.0 nE cm^{-2} s^{-1} (400–700 nm) at the beginning and declined 10% by the end of the experiment. Light was measured with a Bemar Q meter equipped with a Zeta sensor. The light and dark periods were 16 and 8 hours, respectively, with an abrupt change.

The air temperature over each plant throughout the experiment was 25°C $\pm$ 1°C in the light and 20°C $\pm$ 1°C in the dark, as sensed with a shielded 24-gauge thermocouple. Soil temperature at the center of the containers was 20°C $\pm$ 3°C, as sensed with a Proban thermistor. The relative humidity was 65% $\pm$ 8% during the day and 73% $\pm$ 5% at night, as measured with an aspirated psychrometer. The air flow up through the plants was 30 m/min at the top of the plant canopy, as measured with a Windo Model 12 hot-wire anemometer. Fresh makeup air (0.6 m³/min) was provided, and CO_2 was monitored with Manbek infrared gas analyzer and remained above 300 ppm during the light period.

Plants were grown in peat vermiculite: 1:1, v/v in 1-liter white cylindrical polyethylene containers 15 cm in diameter. Plants were placed in the chamber in a randomized block arrangement and irrigated for 5 min every 6 hours with 100 ml of ASHS nutrient solution per container.

*Fictitious brand names are used throughout this sample.

Appendix

Manufacturers of Growth Chambers

While it is impractical to provide a complete list of manufacturers and dealers, this partial list is furnished for your information, with the understanding that no discrimination is intended, and no guarantee of reliability implied. D. T. KRIZEK.

In the United States

Controlled Environments, Inc.
601 Stutsman Street
P.O. Box 347
Pembina, North Dakota 58271
Tel. (204) 786-6451

Environaire System, Inc.
P.O. Box 401
East Longmeadow, Massachusetts 01028
Tel. (413) 525-4336

Environmental Growth Chambers
P.O. Box 407
Chagrin Falls, Ohio 44022
Tel. (216) 247-5100

Forma Scientific, Inc.
Box 649
Marietta, Ohio 45750
Tel. (614) 373-4763

Hotpack Corporation
Hotpack Growth Chambers
5086A Cottman Avenue & Melrose Street
Philadelphia, Pennsylvania 19135
Tel. (215) 333-1700

Instrumentation Specialties Co., Inc. (ISCO)
Environmental Growth Chamber Division
Building 978, Lincoln Air Park West
Lincoln, Nebraska 68524
Tel. (402) 799-2441

Lab-Line Instruments, Inc.
Lab-Line Plaza
15th and Bloomingdale Avenues
Melrose Park, Illinois 60160
Tel. (312) 345-7400

National Appliance Company
P.O. Box 23008
10855 S.W. Greenburg Road
Portland, Oregon 97223
Tel. (503) 639-3161

Parce Engineering Company
P.O. Box 2366
900 W. Van Buren
Harlingen, Texas 78550
Tel. (512) 423-2513

Percival Manufacturing Company
P.O. Box 249
1805 East 4th Street
Boone, Iowa 50036
Tel. (515) 432-6501

Puffer-Hubbard Refrigeration
850 East Jackson
Grand Haven, Michigan 49417

Scientific Systems Corporation
9020 South Choctaw
Baton Rouge, Louisiana 70815
Tel. (504) 926-6950

Sherer Environmental Division
Kysor Industrial Corporation
Environmental Dept.
910 West Industrial Road
Marshall, Michigan 49068
Tel. (616) 781-3911

Outside the United States

Brown Boveri-York
Kalte-und Klimatechnik GmbH
Postfach 346
6800 Mannheim 1
West Germany

Cie Climatechnique
67, rue Morat 68
Colmar
France

Coldstream Limited
1855 Sargent Avenue
Winnipeg 21
Manitoba R 3H OE3
Canada

Colmat Environmental Systems and Controls, Limited
961 West First Street
North Vancouver, British Columbia
Canada

Controlled Environment Limited
661 Century Street
Winnipeg R 3H OL9
Canada

Controlled Environment Limited
20–21 St. Dunstan's Hill
London, EC3R 8PH
  England

Facis S. A.
85, rue Chaptal
92, Levallois
France

Fisons Scientific Apparatus Limited
Bishop Meadow Road
Loughborough, Leic.
England

Froilabo-Sogev.
3 et 5, Bd de Levallois
92, Neuilly-sur-Seine
France

Koito Industries Limited
Environmental Control Division
1-16-6 Miyamae
Suginami-Ku
Tokyo
Japan

Le Materiel Physico-Chimique (Flam et Cie)
B.P.n°4
25, rue Robert Schumann
93, Neuilly sur Marne
France

L'Humidifere Crapez
79, Vales
France

Messrs. Nessi Brothers et Cie
43, rue de la Vanne
92, Montrouge
France

Realis
30, rue Etienne Dolet
94, Villejuif
France

Rubarth and Company
Fabrik med. und elecktr. Apparate
Ikarusallee 2
D 3000 Hannover
West Germany

Sapratin-Environnement
30, rue Raspail
95, Argenteuil
France

Temperature Control Limited
P.O. Box 22013
Otahuhu
Auckland
New Zealand

Thermotechnique Loubriat
9, Place A. Rijckmans
5000 Namur
Belgium

Ernst Votsch
Kalte-and Klimatechnik
KG Box 40
7426 Frommern (Wurth)
West Germany

Karl Weiss
Fabrik Elektro-Physikal-Gerate
D-6301 Lindenstruth
Giessen
West Germany

Index

A GROWTH CHAMBER MANUAL

Designed by Gary Gore.
Composed by Imperial Litho/Graphics,
in 10 point VIP Primer, 2 points leaded,
with display lines in Helvetica and Helvetica Bold.
Printed offset by Vail-Ballou Press on
Warren's No. 66 text, 50 pound basis.
Bound by Vail-Ballou Press
in Joanna book cloth
and stamped in All Purpose foil.

Library of Congress Cataloging in Publication Data
(For library cataloging purposes only)

Main entry under title:
A Growth chamber manual.

  Includes bibliographies and index.
  1. Growth cabinets and rooms—Handbooks, manuals, etc. I. Langhans,
Robert W., 1929-
QK715.G76   635.9'8   77-90906
ISBN 0-8014-1169-6